Exam Success

IET Wiring Regulations 2382-12

Paul Cook and Jonathan Elliott

City & Guilds is a trademark of the City and
Guilds of London Institute
The IET is an imprint of the Institution of
Engineering and Technology

ISBN: 978 0 85193 232 3

First edition published 2006 as *Exam Success:
IEE Wiring Regulations 2381*
Second edition published 2008 as *Exam
Success: IEE Wiring Regulations 2382-10*
This edition published 2012

Every effort has been made to ensure that the
information contained in this publication is
true and correct at the time of going to press.
However, examination products and services
are subject to continuous development and
improvement and the right is reserved to
change products and services from time
to time. While the author, publishers and
contributors believe that the information
and guidance given in this work is correct,
all parties must rely upon their own skill and
judgement when making use of it. Neither the
author, the publishers nor any contributor
assume any liability to anyone for any loss or
damage caused by any error or omission in the
work, whether such error or omission is the
result of negligence or any other cause. Where
reference is made to legislation it is not to be
considered as legal advice. Any and all such
liability is disclaimed.

Cover and book design by CDT Design Ltd
Typeset by Cenveo Publisher Services
Printed in the UK by Swallowtail Print

With thanks to Brian Scaddan

Exam Success

IET Wiring Regulations 2382-12

Paul Cook and Jonathan Elliott

**City & Guilds Level 3 Award in Requirements for Electrical
Installations BS 7671: 2008 (2011)**

City & Guilds is the UK's leading provider of vocational qualifications, offering over 500 awards across a wide range of industries, and progressing from entry level to the highest levels of professional achievement. With over 8500 centres in 100 countries, City & Guilds is recognised by employers worldwide for providing qualifications that offer proof of the skills they need to get the job done.

For publication enquiries:
City & Guilds
1 Giltspur Street
London EC1A 9DD
T +44 (0)844 543 0000
Email centresupport@cityandguilds.com
www.cityandguildsbookshop.com

Queries or feedback on any of our publications should be sent to:
publishingfeedback@cityandguilds.com.

The Institution of Engineering and Technology is the new institution formed by the joining together of the IEE (The Institution of Electrical Engineers) and the IIE (The Institution of Incorporated Engineers). The new institution is the inheritor of the IEE brand and all its products and services, including the IET Wiring Regulations (BS 7671) and supporting material.

For publication enquiries:
The Institution of Engineering and Technology
PO Box 96
Stevenage
SG1 2SD, UK
T +44 (0)1438 767 328
Email sales@theiet.org
www.theiet.org.
www.electrical.theiet.org/books/
regulations/17th-edition-amd1.cfm

Contents

Introduction

How to use this book

This book has been written as a study aid for the City & Guilds Level 3 Award in Requirements for Electrical Installations BS 7671: 2008 (2011) (2382-12). It sets out methods of studying, offers advice on exam preparation and provides details of the scope and structure of the examination, alongside sample questions with fully worked-through answers. Used as a study guide for exam preparation and practice, it will help you to reinforce and test your existing knowledge, and will give you guidelines and advice about sitting the exam. You should try to answer the sample test questions under exam conditions (or as close as you can get) and then review all of your answers. This will help you to become familiar with the types of question that might be asked in the exam and also give you an idea of how to pace yourself in order to complete all questions comfortably within the time limit. This book cannot guarantee a positive exam result, but it can play an important role in your overall revision programme, enabling you to focus your preparation and approach the exam with confidence.

IET Wiring Regulations, 17th Edition, BS 7671:2008 Amendment No 1

You will need a copy of the *IET Wiring Regulations, 17th Edition, BS 7671:2008 Amendment No 1* to be able to answer the sample questions and in order to revise for the examination. The *IET Wiring Regulations* is called, in full, *BS 7671:2008(2011) Requirements for Electrical Installations (IET Wiring Regulations, 17th Edition, incorporating Amendment No 1)*. The IET Wiring Regulations is the national standard used in the UK for the safe design and erection of electrical installations.

City & Guilds Level 3 Award in Requirements for Electrical Installations BS 7671: 2008 (2011) (2382-12)

If you are an electrical installer with relevant experience, you need to achieve this Level 3 Award to show you have a sound working knowledge of the format, content and application of the current edition of the wiring regulations, the BS 7671:2008 (2011) Requirements for Electrical Installations (IET Wiring Regulations, 17th Edition, incorporating Amendment No 1). This award is also useful for electricians who have studied electrical installation practice in other countries, and are familiar with either the International Standards in the IEC 60364 Electrical Installations of Buildings series, or European Standard Series HD 0364 Electrical Installations of Buildings.

This qualification includes all new material and updates to the *IET Wiring Regulations*. If you already have a certificate for the Wiring Regulations but it was awarded prior to the Seventeenth Edition in 2008, then you will need to take the 2382-12 exam.

The syllabus covers topics such as protection for safety, selection and erection of equipment, and inspection and testing to meet the standards of the *IET Wiring Regulations*.

Finding a centre

In order to take the exam, you must register at an approved City & Guilds centre. You can find your nearest centre by looking up the qualification number 2382-12 at www.cityandguilds.com.

At each centre, the Local Examinations Secretary will enter you for the award, collect your fees, arrange for your assessment to take place and correspond with City & Guilds on your behalf. The Local Examinations Secretary also receives all of your certificates and correspondence from City & Guilds. Most centres will require you to attend a course of learning before entering you for the examination. These can vary between week-long intensive courses and courses running across a term, once a week for two to three hours.

Awarding and reporting

When you complete the City & Guilds 2382-12 Certificate online examination, you will be given your provisional results, as well as a breakdown of your performance in the various areas of the examination. This is a useful diagnostic tool if you fail the exam, as it enables you to identify your individual strengths and weaknesses across the different topics.

A Certificate is issued automatically when you have been successful in the assessment, but it will not indicate a grade or percentage pass. Your centre will receive your Notification of Candidate's Results and Certificate. Any correspondence is conducted through the centre. The centre will also receive consolidated results lists detailing the performance of all candidates entered. If you have particular requirements that will affect your ability to attend and take the examination, then your centre should refer to City & Guilds policy document 'Access to Assessment: Candidates with Particular Requirements'.

The exam

The exam

The exam

The examination has a multiple-choice format, with 60 questions, which you will have two hours to answer. The test is offered on **e-volve**, a simple online service that does not require strong IT skills. **e-volve** uses a bank of questions set and approved by City & Guilds. Each candidate receives randomised questions, so no two candidates will sit exactly the same test.

The exam always follows a set structure, based on the seven parts of the *IET Wiring Regulations* plus the appendices. The table below outlines the sections of the exam and the number of questions in each section. It also shows the weighting – so you can see how important each section is in determining your final score.

Section	Topic		% weighting	No of questions
1	Part 1	Scope, object and fundamental principles	7	4
2	Part 2	Definitions	3	2
3	Part 3	Assessment of general characteristics	10	6
4	Part 4	Protection for safety	25	15
5	Part 5	Selection and erection of equipment	23	14
6	Part 6	Inspection and testing	7	4
7	Part 7	Special installations or locations	17	10
8	Use of appendices		8	5
	Total		100	60

Sitting a City & Guilds online examination

The test will be taken under usual exam conditions. If you leave the exam room unaccompanied before the end of the test period, you may not be allowed to come back into the room.

Introduction

When you take a City & Guilds exam online, you can go through an introduction to familiarise yourself with the system. Once you are logged on to take the exam, the first screen will give you the chance to look through a set of introductory screens that show how the exam will be presented, how to get help, how to move between different screens and how to mark questions that you might want to return to later.

Notes

The sample questions in the introduction will be unrelated to the exam you are taking. The time it takes to look through the introduction is not included in the exam time, so you can go through it in your own time. The exam will only start once you have completed or skipped the introduction.

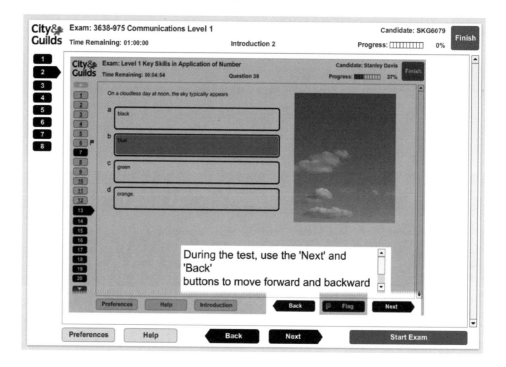

Use the 'Next' and 'Back' buttons to navigate through the introduction.

You can quit the introduction and start the test at any time by clicking on the 'Start Exam' button. The introduction will still be available for you to go back to at any time during the exam, although once you have started the exam the clock timer will continue to move forwards and cannot be paused.

The 'Preferences' button allows you to change the colour scheme of the exam. The different options are designed to improve accessibility for candidates with a visual impairment. Simply select from the list the option you find easiest to read and click 'OK'. The 'Restore defaults' button will set the display back to the default style.

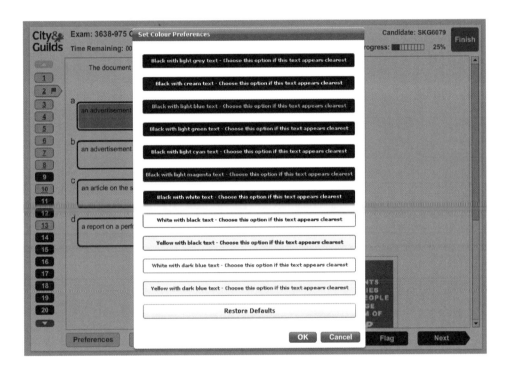

The 'Preferences' button can be used before the exam begins (ie during the introduction) or at any point during the exam. However, the timer will continue to move on if you look at the preferences and change them during the exam. To maximise your available exam time, we advise that you review your preferences before you start the exam.

The 'Help' button will launch a tutorial that explains how to use the system. Select the help option you need from the dropdown menu at the bottom, or move backwards and forwards with the 'Back' and 'Next' buttons.

The 'Help' button can be used before the exam begins (ie during the introduction) or at any point during the exam. However, the timer will continue to move on if you look at the help options during the exam. To maximise your available exam time, we advise that you review the Help screens before you start the exam.

Once you have finished with the introduction you can click the 'Start Exam' button.

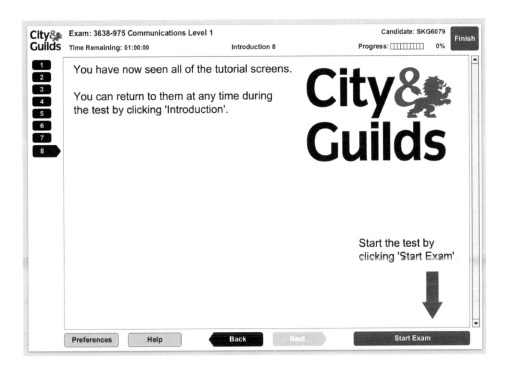

You will then see the first question displayed and the timer will start.

The exam

This is a typical screen that you will see in the exam.

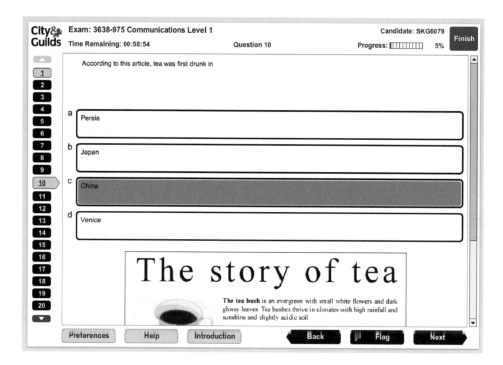

At the top of the screen you can see your progress (on the right-hand side) and the exam time remaining (on the left-hand side). A notification will be displayed when you have 30, 15 and 5 minutes left.

The bar on the left-hand side shows the question numbers. You can progress through the test using the 'Next' and 'Back' buttons at the bottom of the page, or you can go straight to a specific question by clicking on that question number on the bar on the left.

You may flag a question to come back to it later by clicking on the 'Flag' button.

Some of the questions in the test may be accompanied by an exhibit. This may be an article or table for you to scroll through, or a photograph or other illustration. The question will tell you whether you will need to click on the 'Document' or 'Scenario' button to view these items.

When you reach the final question and click 'Next', or if you click 'Finish' at any time, you will have the option to finish the test. If you have unanswered or flagged questions, the system will warn you. Click 'Cancel' to go back to the test and review your answers. Click 'Finish' to proceed.

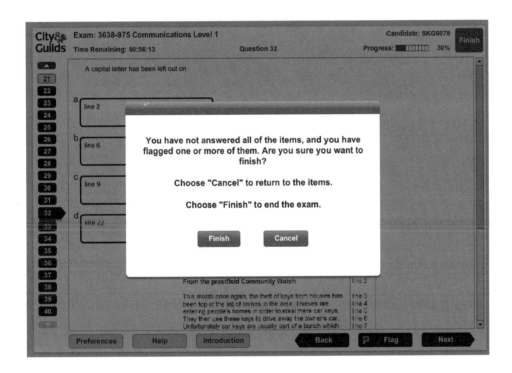

A final warning will then be displayed. If you click the 'Exit' button the test will finish and you will not be able to go back.

You will then be prompted to close the application.

Your Invigilator will be able to give you feedback on your result as soon as the test is finished (provided you are sitting a multiple choice exam that does not need to be marked by an examiner).

Frequently asked questions

When can I sit the paper?
You can sit the exam at any time, as there are no set exam dates. You may need to check with your centre when it is able to hold exam sessions.

Can I use any reference books in the test?
Yes, you can use a copy of the IET Wiring Regulations, 17th Edition, BS 7671:2008 Amendment No 1.

How many different parts of the test are there?
There are 60 questions, which cover eight different sections of the syllabus.

Do I have a time limit for taking the test?
You have two hours to complete the test.

Do I need to be good at IT to do the test online?
No, the system is really easy to use, and you can practise before doing the test.

What happens if the computer crashes in the middle of my test?
This is unlikely, because of the way the system has been designed. If there is some kind of power or system failure, then your answers will be saved and you can continue on another machine if necessary.

Can people hack into the system and cheat?
There are lots of levels of security built into the system to ensure its safety. Also, each person gets a different set of questions, which makes it very difficult to cheat.

Can I change my answer?
Yes, you can change your answers quickly, easily and clearly at any time in the test up to the point where you end the exam. With any answers you feel less confident about, you can click the 'Flag' button, which means you can review these questions before you end the test.

How do I know how long I've got left to complete the test?
You can check the time remaining at any point during the exam by clicking on the 'Clock' icon in the tool bar. The time remaining will come up on the bottom right corner of the screen.

Is there only one correct A, B, C or D answer to multiple-choice questions?
Yes.

What happens if I don't answer all of the questions?
You should attempt to answer all of the questions. If you find a question difficult, mark it using the 'Flag' button and return to it later.

What grades of pass are there?
A Pass or a Fail.

When can I resit the test if I fail?
You can resit the exam at any time, and as soon as you and your tutor decide it is right for you, subject to the availability of the online examination.

Notes

Exam content

To help you to fully understand the exam content, this chapter is divided into the eight sections of the exam, which are in turn mapped to the seven parts and the appendices of BS 7671 (the *IET Wiring Regulations*). The requirements of the examination are listed and references made to the relevant parts of BS 7671. One of the most useful parts of BS 7671 is its very comprehensive index. If you don't know where to look, use the index.

Section 1

Part 1 – Understand the scope, object and fundamental principles of BS 7671

There are four questions on Part 1, which represent 7 per cent of the mark. Part 1 of BS 7671 contains Chapters 11, 12 and 13 and is the essence of the standard. It includes the scope, an important section as any installation for which requirements are being sought must be confirmed as being within the scope of BS 7671. It also includes the fundamental principles (Chapter 13). The rest of BS 7671, that is Parts 2 to 7, supports Part 1; Part 2 provides definitions of the terminology used throughout the standard and Parts 3 to 7 provide technical requirements intended to ensure that the electrical installation conforms with the fundamental principles of Chapter 13 of Part 1.

It is to be noted that departures from Parts 3 to 7 are allowed (Regulation 120.3), providing there is still compliance with Part 1. Regulation 133.5 allows the use of new materials and inventions not anticipated when the Regulations were published. However, departures from Parts 3 to 7, including the use of new materials and inventions, must not result in a degree of safety any less than that required by Part 1. Departures from Parts 3 to 7 must be recorded on the installation certificate.

To show that you are conversant with BS 7671, you are required to be able to:
1.1 Identify the scope of BS 7671
1.2 Identify the object of BS 7671
1.3 Identify the fundamental principles of BS 7671.

Section 2

Part 2 – Understand the definitions used within BS 7671

There are two questions on Part 2, which represent 3 per cent of the mark. There are approximately 14 pages dedicated to the definitions used by the Regulations. Definitions that are particularly important, as their meanings in BS 7671 are often not the same as those in common English usage, include:

> **Exposed-conductive-part.** A conductive part of equipment which can be touched and which is not normally live, but which can become live under fault conditions.
>
> **Extraneous-conductive-part.** A conductive part liable to introduce a potential, generally Earth potential, and not forming part of the electrical installation.
>
> **Electrical equipment** (abbr: *Equipment*). Any item for such purposes as generation, conversion, transmission, distribution or utilisation of electrical energy, such as machines, transformers, apparatus, measuring instruments, protective devices, wiring systems, accessories, appliances and luminaires. (Be sure to note that 'equipment' is the abbreviation for 'electrical equipment'.)

To show that you are conversant with BS 7671, you are required to be able to:

2.1   Interpret the definitions used within BS 7671
2.2   Relate the definitions to the regulations and appendices of BS 7671.

Section 3

Part 3 – Understand how to assess the general characteristics of electrical installations

There are six questions on Part 3, which represent 10 per cent of the mark. Part 3 requires a general assessment of the characteristics of the proposed installation. It also requires that the nature or characteristics of the supply be determined, as this plays an important part in the design of the installation. Consideration needs to be given to safety services and standby supplies. This part also includes requirements for the installation circuit arrangements.

Part 3 also includes useful technical information, such as the definitions of types of system earthing, eg '**TN-S system.** A system having separate neutral and protective conductors throughout the system.' A black and white drawing is provided as Fig 3.8 in Part 3 of the Regulations; for a colour drawing see page 22.

TN-S system

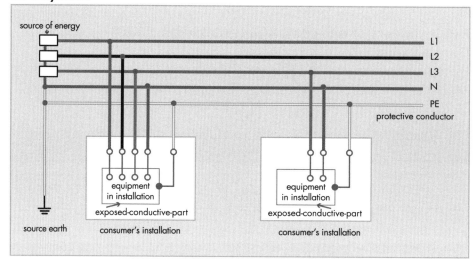

Separate neutral and protective conductors throughout the system.

The protective conductor (PE) is the metallic covering of the cable supplying the installations or a separate conductor.

All exposed-conductive-parts of an installation are connected to this protective conductor via the main earthing terminal of the installation.

To show that you are conversant with BS 7671, you are required to be able to:

3.1   Interpret the requirements of assessing the general characteristics of electrical installations within the scope of BS 7671.

Section 4

Part 4 – Understand the requirements of protection for safety for electrical installations

There are 15 questions on Part 4, which represent 25 per cent of the mark. Part 4 contains some of the key chapters of BS 7671: Chapter 41, Chapter 42, Chapter 43 and Chapter 44.

To show that you are conversant with BS 7671, you are required to be able to:

4.1   Identify the requirements of protection for safety within the scope of BS 7671

4.2   Interpret how this applies to electrical installations within the scope of BS 7671 to include:
   – protection against electric shock
   – protection against thermal effects
   – protection against overcurrent
   – protection against voltage disturbances and electromagnetic disturbances.

Section 5

Part 5 – Understand the requirements for selection and erection of equipment for electrical installations

There are 14 questions on Part 5, which represent 23 per cent of the mark.

Part 5 includes requirements for voltage drop, warning notices and identification of conductors, as well as requirements for isolation and switching, earthing and bonding arrangements, protective conductors, lighting installations and generating sets.

To show that you are conversant with BS 7671, you are required to be able to:

5.1 Identify the requirements for selecting and erecting equipment, within the scope of BS 7671

5.2 Interpret how this applies to electrical installations within the scope of BS 7671 to include:
- common rules
- selection and erection of wiring systems
- protection, isolation, switching, control and monitoring
- earthing arrangements and protective conductors
- other equipment
- safety services.

Section 6

Part 6 – Understand the requirements of inspection and testing of electrical installations

There are four questions on Part 6, which represent 7 per cent of the mark.

To show that you are conversant with BS 7671, you are required to be able to:

6.1 Identify the requirements for inspection and testing

6.2 Interpret how this applies to electrical installations.

Section 7

Part 7 – Understand the requirements of special installations or locations as identified in BS 7671

There are ten questions on Part 7, which represent 17 per cent of the mark. Part 7 includes additional requirements or altered requirements to those in Parts 3 to 6. The requirements of the rest of BS 7671 apply to the special locations, unless they are amended or altered by the particular requirements in Part 7.

Notes

To show that you are conversant with BS 7671, you are required to be able to:

7.1   Identify the requirements for special installations and locations

7.2   Interpret how these affect the general requirements of the regulations.

Section 8

There are five questions on the Appendices of BS 7671, which represent 8 per cent of the mark.

To show that you are conversant with BS 7671, you are required to be able to:

8.1   Identify the information in the appendices of BS 7671

8.2   Specify how the information contained in the appendices is used to support electrical installation activities.

Tips from the examiner

The following tips are intended to aid confident test performance. Some are more general and would apply to most exams. Others are more specific, either because of the format of this test (multiple choice) or the nature of the subject.

✔ If you rarely use a computer, try to get some practice beforehand. You need to be able to use a mouse to move a cursor arrow around a computer screen, as you will use the cursor to click on the correct answer in the exam.

✔ Take a copy of *BS 7671:2008(2011) Requirements for Electrical Installations (IET Wiring Regulations, 17th Edition, incorporating Amendment No 1)* into the exam. Take the time to familiarise yourself with the structure and content of this publication.

✔ Make the most of the course you will attend before taking the exam. Try to attend all sessions and be prepared to devote time outside the class to revise for the exam.

✔ On the day of the exam, allow plenty of time for travel to the centre and arrive at the place of the exam at least ten minutes before it's due to start so that you have time to relax and get into the right frame of mind.

✔ Listen carefully to the instructions given by the invigilator.

✔ Read the question and every answer before making your selection. Do not rush – there should be plenty of time to answer all the questions.

✔ Look at the exhibits where instructed. Remember, an exhibit supplies you with information that is required to answer the question.

✔ Attempt to answer all the questions. If a question is not answered, it is marked as wrong. Making an educated guess improves your chances of choosing the correct answer. Remember, if you don't select an answer, you will definitely get no marks for that question.

✔ The order of the exam questions follows the order of the *IET Wiring Regulations* publication. Therefore, look for the answers to early questions at the front of the book and progress through it as you work through the exam questions.

Notes

✔ Don't worry about answering the questions in the order in which they appear in the exam. Choose the 'Flag' option on the tool bar to annotate the questions you want to come back to. If you spend too much time on questions early on, you may not have time to answer the later questions, even though you know the answers.

✔ You should bring a basic, non-programmable calculator to the exam. You might find it helpful to bring one with square-root ($\sqrt{}$) and square (x^2) functions.

✔ If you are having trouble finding the regulation for a particular question, look for the subject in the index of the *IET Wiring Regulations*. Using the index should minimise the time it takes you to find the relevant topic.

✔ It is not recommended that you memorise any of the material presented here in the hope it will come up in the exam. The exam questions featured in this book will help you to gauge the kinds of questions that might be asked. It is highly unlikely you will be asked any identical questions in the exam, but you may see variations on certain themes.

Exam practice 1

Exam practice 1

Sample test 1

The sample test below has 60 questions, the same number as the online exam, and its structure follows that of the online exam. The test appears first without answers, so you can use it as a mock exam. It is then repeated with worked-through answers and extracts from the *IET Wiring Regulations*. Finally, there is an answer key for easy reference.

Answer the questions by filling in the circle next to your chosen option.

Section 1

1 BS 7671 applies to

- ○ a   equipment on board ships
- ○ b   lightning protection of buildings
- ○ c   lift installations
- ○ d   prefabricated buildings.

2 BS 7671 provides requirements for safety against the risk of

- ○ a   electric shock on an aircraft
- ○ b   shock currents on board ships
- ○ c   fire on offshore installations
- ○ d   shock currents in electrical installations.

3 Which one of the following is <u>not</u> a statutory regulation?

- ○ a   Electricity at Work Regulations 1989 as amended.
- ○ b   The Supply of Machinery (Safety) Regulations 1992 as amended.
- ○ c   Requirements for Electrical Installations (BS 7671).
- ○ d   Agricultural (Stationary Machinery) Regulations.

4 BS 7671 identifies that the cross-sectional area of a conductor shall be determined by

- ○ a   the admissible maximum temperature
- ○ b   the nominal voltage
- ○ c   voltage tolerances
- ○ d   the earthing system.

Section 2

5   A corridor containing supporting structures for cables and joints and/or other elements of wiring systems, the dimensions of which allow persons to pass freely throughout the entire length, is known as

○ a   an access pathway
○ b   a cable tunnel
○ c   an access throughway
○ d   cable ducting.

6   The algebraic sum of the currents in the live conductors of a circuit at a point in the electrical installation is known as the

○ a   residual current
○ b   harmonic current
○ c   line current
○ d   neutral current.

Section 3

7   The current-carrying conductor arrangement shown in the figure below is

○ a   single-phase 3-wire phase angle 0°
○ b   two-phase 3-wire phase angle 90°
○ c   three-phase 3-wire star connected
○ d   three-phase 3-wire delta connected.

8   With reference to the nature of the supply, which one of the following can be determined by calculation, enquiry or measurement?

○ a   The maximum demand of the installation.
○ b   The rating of the circuit protective device.
○ c   The prospective short-circuit current at the origin of the installation.
○ d   The csa of the tails.

9 Every installation should be divided into individual circuits to

○ a   prevent faults developing
○ b   provide protection against electric shock
○ c   minimize inconvenience in the event of a fault
○ d   ease installation.

10 Diversity may be taken into account when considering

○ a   maximum demand of the installation
○ b   a TN-C-S system
○ c   the prospective short-circuit fault current
○ d   the number of final circuits.

11 Which one of the following is <u>not</u> a characteristic of the supply?

○ a   The nature of the current and frequency.
○ b   The earth fault loop impedance external to the installation.
○ c   Main switch current rating.
○ d   The nominal voltage.

12 BS 7671 requires designers to take into account the frequency and quality of maintenance an installation can reasonably be expected to receive when

○ a   assessing staff numbers
○ b   inspecting and testing
○ c   selecting staff
○ d   specifying or selecting equipment reliability.

Section 4

13 Which one of the following is <u>not</u> part of the requirements for fault protection?

○ a   Protective earthing.
○ b   Protective equipotential bonding.
○ c   Automatic disconnection.
○ d   Protection by insulation of live parts.

14 The maximum disconnection time for a TN system with a nominal voltage of 400 V a.c. to Earth is

○ a   0.2 second
○ b   0.4 second
○ c   0.5 second
○ d   5 seconds.

15 For a TT system, which one of the following conditions should be fulfilled for each circuit protected by an RCD?

○ a $R_A \times I_{\Delta n} \leq 50$ V
○ b $R_A \times I_{\Delta n} \geq 50$ V
○ c $R_A \times I_d \leq 50$ V
○ d $R_A \times I_d \geq 50$ V

16 A 32 A type B circuit-breaker is used to give a disconnection time of 5 seconds in a reduced low voltage system with a nominal voltage to Earth (U_O) of 55 V. What is the maximum value of earth fault loop impedance (Z_S)?

○ a   0.44 Ω.
○ b   0.34 Ω.
○ c   0.17 Ω.
○ d   0.09 Ω.

17 A SELV source can be derived from which one of the following?

○ a   Double-wound transformer.
○ b   Autotransformer.
○ c   Safety isolating transformer.
○ d   Step-up transformer.

18 Which one of the following <u>cannot</u> be used as basic protection?

○ a   Insulation of live parts.
○ b   Barriers or enclosures.
○ c   Protective earthing and bonding.
○ d   Obstacles.

Notes

19 In order to provide basic protection, a horizontal top surface of a barrier or enclosure that is readily accessible shall provide a minimum degree of protection of

○ a   IPXXA or IP1X
○ b   IPXXB or IP2X
○ c   IPXXC or IP3X
○ d   IPXXD or IP4X.

20 Where arcs, sparks or particles at high temperature may be emitted by fixed equipment in normal service, the equipment shall be

○ a   totally enclosed in arc-resistant material
○ b   protected by a 30 mA RCD
○ c   enclosed to at least IP55
○ d   accessible only by use of a key or tool.

21 Except where otherwise recommended by the manufacturer, spotlights and projectors rated at over 100 W and up to 300 W shall be installed at a minimum distance from combustible materials of

○ a   0.5 m
○ b   0.6 m
○ c   0.8 m
○ d   1.0 m.

22 In locations with increased risks of fire, motors which are automatically or remotely controlled, or which are not continuously supervised, shall be protected against excessive temperature by

○ a   a protective device that is automatically reset
○ b   a protective device with manual reset
○ c   electronic monitoring equipment that resets
○ d   electronic monitoring equipment that restarts the motor.

23 Where particular risks of fire exist, the classification for high density occupation areas with easy conditions of evacuation is

- a   BD1
- b   BD2
- c   BD3
- d   BD4.

24 When considering protection against overload, the symbol for the current ensuring effective operation of the protective device is

- a I_b
- b I_z
- c I_n
- d I_2.

25 For protection against overvoltage a 230 V electricity meter should have an impulse withstand of

- a   6 kV
- b   4 kV
- c   2.5 kV
- d   1.5 kV.

26 What is the impulse category of equipment that is part of the fixed electrical installation and other equipment where a high degree of availability is expected?

- a   I.
- b   II.
- c   III.
- d   IV.

27 Which of the following is <u>not</u> considered as a potential source of electromagnetic disturbance?

- a   Electric motors.
- b   Rectifiers.
- c   Electric fires.
- d   Lifts.

Section 5

28 A durable label to BS 951 with the words 'Safety Electrical Connection – Do Not Remove' is <u>not</u> required at

○ a   the connection of every earthing conductor to an earth electrode
○ b   the point of connection of every bonding conductor to an extraneous-conductive-part
○ c   the main earth terminal, where separate from the main switchgear
○ d   a main earthing bar contained within switchgear.

29 With regards to temporary overvoltages (TOVs), a surge protective device (SPD) should be selected and erected in accordance with the

○ a   ESQCR
○ b   DNO
○ c   requirements of BS EN 60309
○ d   manufacturer's instructions.

30 A firefighter's switch shall be provided in the low voltage circuit supplying exterior electrical installations and interior discharge lighting operating at

○ a   a voltage exceeding low voltage
○ b   low voltage
○ c   voltage band II
○ d   medium voltage.

31 A plug and socket-outlet may be used for switching off for mechanical maintenance as long as it does not have a rating exceeding

○ a   13 A
○ b   16 A
○ c   32 A
○ d   45 A.

32 If an area within an installation undergoes a 10 °C rise in ambient temperature, the effect on the current-carrying capacity of cables will be to

○ a   decrease the value of I_z
○ b   increase the value of I_z
○ c   leave I_z unchanged
○ d   increase the fault current by 10 per cent.

33 A multicore 70 °C thermoplastic cable with 2.5 mm² conductors supplies a single-phase load of 20 A at 230 V a.c. over a distance of 22 metres. The voltage drop in the cable will be

○ a   6 V
○ b   6.6 V
○ c   7.92 V
○ d   9.2 V.

34 Where a 3-core cable, with cores coloured brown, black and grey, is used as a switch wire for two-way or intermediate control, the terminations of the conductors shall be identified using

○ a   red, blue and yellow tape
○ b   black tape only on each core
○ c   brown tape on the black and grey cores
○ d   self-colour tape only.

35 Which one of the following <u>cannot</u> be used as an earth electrode?

○ a   Earth plates.
○ b   Welded reinforcement of concrete embedded in the earth.
○ c   Earth tapes.
○ d   Gas and water utility pipes.

36 An earthing conductor buried in the ground is protected against corrosion by a sheath, but is not protected against mechanical damage. The minimum size copper conductor that may be installed is

○ a   2.5 mm²
○ b   16 mm²
○ c   25 mm²
○ d   50 mm².

Notes

37 Assuming that both the line and protective conductors are of the same material, for a line conductor of 10 mm^2 if the protective conductor is to be selected, the minimum tabulated cross-sectional area of its associated protective conductor is

- ○ a   6 mm^2
- ○ b   10 mm^2
- ○ c   16 mm^2
- ○ d   35 mm^2.

38 A radial final circuit feeding socket-outlets supplying several items of data processing equipment has a total protective conductor current in normal service of 18 mA. This circuit must have a high integrity protective conductor

- ○ a   of cross-sectional area less than 1 mm^2
- ○ b   connected as a ring
- ○ c   controlled by an isolator
- ○ d   enclosed in insulated conduit only.

39 Where an uninterruptible power supply is used it shall comply with

- ○ a   BS 3036
- ○ b   BS 88-3
- ○ c   BS EN 60898 and BS EN 61009
- ○ d   BS EN 62040-1 and BS EN 62040-3.

40 Where an autotransformer is connected to a circuit having a neutral conductor, the common terminal of the winding shall be connected to the

- ○ a   neutral conductor
- ○ b   line conductor
- ○ c   protective conductor
- ○ d   bonding conductor.

41 When selecting wiring systems for safety services, the type of cable that should be used in fire conditions should comply with the test requirements of

○ a   BS 5467
○ b   BS 6231
○ c   IEC 60331-1
○ d   BS EN 50200.

Section 6

42 When carrying out a visual inspection of an electrical installation, which one of the following does <u>not</u> have to be verified?

○ a   The methods of protection against electric shock.
○ b   The electricity supplier.
○ c   The connection of conductors.
○ d   The presence of undervoltage protective devices.

43 Certain information must be made available to persons carrying out inspection and testing of an installation before the testing commences. One such item of information would be

○ a   the name of the client
○ b   the name of the person who designed the installation
○ c   the length of cable runs in the installation
○ d   any circuit or equipment vulnerable to a typical test.

44 An insulation resistance test is to be carried out on a 3-phase 400 V circuit. The test voltage and minimum acceptable reading would be

○ a   250 V a.c. and 0.5 MΩ
○ b   500 V d.c. and 0.5 MΩ
○ c   500 V d.c. and 1 MΩ
○ d   800 V d.c. and 0.5 Ω.

45 The document which must be issued following the periodic inspection and testing of an electrical installation is

○ a   a Periodic Inspection Report
○ b   a Minor Electrical Installation Works Certificate
○ c   an Electrical Installation Condition Report
○ d   an Electrical Installation Certificate.

Notes

Section 7

46 In a room containing a bath, electrical equipment installed in zone 0 shall have a degree of protection of at least

- ○ a   IPX5
- ○ b   IP5X
- ○ c   IP7X
- ○ d   IPX7.

47 A flush downlighter in the ceiling less than 2.25 m height above floor level over a bath will need to fulfil the requirements of

- ○ a   zone 2
- ○ b   zone 1
- ○ c   zone 0
- ○ d   all zones.

48 Which of the following protective measures is permitted in a room containing a bath or shower?

- ○ a   Obstacles.
- ○ b   Placing out of reach.
- ○ c   Automatic disconnection of supply.
- ○ d   Non-conducting location.

49 In a swimming pool or other basin, the metallic covering or sheath of a wiring system in zones 0, 1 or 2 shall

- ○ a   not be used
- ○ b   not be earthed
- ○ c   be connected to the supplementary bonding
- ○ d   have reinforced insulation.

50 The requirements of Section 704 of BS 7671 apply to

- ○ a   cloakrooms
- ○ b   offices
- ○ c   construction and demolition site installations
- ○ d   toilets.

51 In locations where livestock is kept, for all circuits other than socket-outlet circuits, an RCD shall be provided with a rating not exceeding

- a 500 mA
- b 300 mA
- c 100 mA
- d 30 mA.

52 Equipment on a pontoon in a marina that is subject to impact to level AG2 should have a mechanical protection code of

- a IPX4
- b IPX8
- c IP55
- d IK08.

53 Distribution boards for Group 2 medical locations should be clearly labelled and installed

- a remote from Group 2 locations
- b in close proximity to Group 2 locations
- c within adjacent Group 1 locations
- d in close proximity to Group 1 locations.

54 In caravans, each final circuit shall be protected against overcurrent by a device that disconnects

- a all live conductors in that circuit
- b the line conductors only
- c line, neutral and protective conductors
- d the caravan and site supply system.

55 Equipment doors or hinged panels in operating or maintenance gangways must be able to open to a minimum of

- a 20 degrees
- b 30 degrees
- c 60 degrees
- d 90 degrees.

Section 8

56 The requirements of the licensing authority should be ascertained for

- a   installations on caravan parks
- b   temporary supplies
- c   types of earthing system to be used
- d   the design of the installation.

57 I_2 can be greater than 1.45 I_z when the overcurrent device

- a   is providing overload protection
- b   is a circuit-breaker
- c   is rated at over 100 A
- d   is providing fault current protection only.

58 External influences coded BE are classified as

- a   nature of processed or stored materials
- b   conditions of evacuation in an emergency
- c   movement of air
- d   capability of persons.

59 The maximum permitted voltage drop for a 230 V lighting circuit supplied from a public low voltage system is

- a   3%
- b   5%
- c   6%
- d   8%.

60 An Electrical Installation Certificate should be made out and signed or otherwise authenticated by

- a   a responsible person
- b   an authorised person
- c   a competent person
- d   a qualified person.

Questions and answers

The questions in sample test 1 are repeated below with worked-through answers and extracts from the *IET Wiring Regulations* where appropriate.

Section 1

1 BS 7671 applies to

○ a   equipment on board ships
○ b   lightning protection of buildings
○ c   lift installations
◉ d   prefabricated buildings.

Answer d
See Part 1, Chapter 11, Scope. Option d, prefabricated buildings, is listed in Regulation 110.1.1, item (vi) as being within the scope. Options a, b and c – equipment on board ships, lightning protection of buildings and lift installations – are listed in Regulation 110.2 as being excluded from the scope of the Regulations.

2 BS 7671 provides requirements for safety against the risk of

○ a   electric shock on an aircraft
○ b   shock currents on board ships
○ c   fire on offshore installations
◉ d   shock currents in electrical installations.

Answer d
Regulation 110.2, Exclusions from scope, advises that options a, b and c are outside the scope.

Regulation 131.1 (i) shock currents in Chapter 13, Fundamental Principles, confirms that the requirements of BS 7671 are intended to provide for the safety of persons and livestock against the risk of injury from shock currents.

3 Which one of the following is <u>not</u> a statutory regulation?

○ a   Electricity at Work Regulations 1989 as amended.
○ b   The Supply of Machinery (Safety) Regulations 1992 as amended.
◉ c   Requirements for Electrical Installations (BS 7671).
○ d   Agricultural (Stationary Machinery) Regulations.

Notes

Answer c

Despite having the subtitle IET Wiring Regulations, Requirements for Electrical Installations, BS 7671:2008(2011) is non-statutory; see Regulation 114.1.

4 **BS 7671 identifies that the cross-sectional area of a conductor shall be determined by**

- ⊙ a   the admissible maximum temperature
- ○ b   the nominal voltage
- ○ c   voltage tolerances
- ○ d   the earthing system.

Answer a

Design is considered in Section 132, and Regulation 132.6 (Cross-sectional area of conductors) is particularly relevant. Option a – 'the admissible maximum temperature' – is listed in 132.6 (i) as being a factor in determining the cross-sectional area. Voltage (options b and c) may affect the type of wiring (see Regulation 132.7). Option d may affect the method of protection against electric shock (see Chapter 41).

132.6 Cross-sectional area of conductors

The cross-sectional area of conductors shall be determined for both normal operating conditions and, where appropriate, for fault conditions according to:

(i)   **the admissible maximum temperature**
(ii)   the admissable voltage drop
(iii)   the electromechanical stresses likely to occur due to short-circuit and earth fault currents
(iv)   other mechanical stresses to which the conductors are likely to be exposed
(v)   the maximum impedance for correct operation of short-circuit and earth fault protection
(vi)   the method of installation
(vii)   harmonics
(viii) thermal insulation.

Section 2

5   A corridor containing supporting structures for cables and joints and/or other elements of wiring systems, the dimensions of which allow persons to pass freely throughout the entire length, is known as

- ○ a   an access pathway
- ◉ b   a cable tunnel
- ○ c   an access throughway
- ○ d   cable ducting.

Answer b

Options a and c are not defined in BS 7671. Options b and d are both defined in Part 2 Definitions, which show that option b is correct.

Cable ducting. An enclosure of metal or insulating material, other than conduit or cable trunking, intended for the protection of cables which are drawn in after erection of the ducting.

Cable tunnel. A corridor containing supporting structures for cables and joints and/or other elements of wiring systems and whose dimensions allow persons to pass freely throughout the entire length.

6   The algebraic sum of the currents in the live conductors of a circuit at a point in the electrical installation is known as the

- ◉ a   residual current
- ○ b   harmonic current
- ○ c   line current
- ○ d   neutral current.

Answer a

The answer is found in Part 2 Definitions, where option a is defined. Option b, harmonic currents, have frequencies that are multiples of the fundamental. Options c and d, line and neutral currents, are the currents flowing in line and neutral conductors. They are not defined.

Residual current. Algebraic sum of the currents in the live conductors of a circuit at a point in the electrical installation.

Notes

Section 3

7 **The current-carrying conductor arrangement shown in the figure below is**

- ○ a single-phase 3-wire phase angle 0°
- ○ b two-phase 3-wire phase angle 90°
- ◉ c three-phase 3-wire star connected
- ○ d three-phase 3-wire delta connected.

Answer c

The answer to this will be found in Part 3 Assessment of General Characteristics Regulation 312.1.1, 'Current-carrying conductors in a.c. circuits' Fig 3.4.

8 **With reference to the nature of the supply, which one of the following can be determined by calculation, enquiry or measurement?**

- ○ a The maximum demand of the installation.
- ○ b The rating of the circuit protective device.
- ◉ c The prospective short-circuit current at the origin of the installation.
- ○ d The csa of the tails.

Answer c

Regulation 313.1 allows that certain of the supply characteristics may be determined either by enquiry of the electricity distributor/supplier or by measurement on-site or, in some circumstances such as a proposed distribution system, by calculation. The prospective short-circuit current at the origin is normally determined by enquiry of the distributor, but could also be measured on-site in some circumstances. Options a, b and d are not listed as characteristics of the supply.

9 **Every installation should be divided into individual circuits to**

- ○ a prevent faults developing
- ○ b provide protection against electric shock
- ◉ c minimize inconvenience in the event of a fault
- ○ d ease installation.

Answer c

Regulation 314.1, item (i), shows that the answer is option c. Dividing an installation into individual circuits does not prevent faults developing (option a) or provide protection against electric shock (option b). It may ease installation (option d), but that is not the reason that BS 7671 requires it.

314.1 Every installation shall be divided into circuits, as necessary, to:
(i) **avoid danger and minimize inconvenience in the event of a fault**
(ii) facilitate safe inspection, testing and maintenance (see also Section 537)
(iii) take account of hazards that may arise from the failure of a single circuit such as a lighting circuit
(iv) reduce the possibility of unwanted tripping of RCDs due to excessive protective conductor (PE) currents not due to a fault
(v) mitigate the effects of electromagnetic disturbances (see also Chapter 44)
(vi) prevent the indirect energizing of a circuit intended to be isolated.

10 Diversity may be taken into account when considering

- ⦿ a   maximum demand of the installation
- ◯ b   a TN-C-S system
- ◯ c   the prospective short-circuit fault current
- ◯ d   the number of final circuits.

Answer a

Diversity between the instant of peak demand of the various loads of an installation may mean that the maximum demand at any time is less than the simple sum of the loads (see Regulation 311.1). Diversity may be taken into account when determining maximum demands, otherwise switchgear and cables may be oversized.

311.1 For economic and reliable design of an installation within thermal limits and admissable voltage drop, the maximum demand shall be determined. In determining the maximum demand of an installation or part thereof, diversity may be taken into account.

11 Which one of the following is <u>not</u> a characteristic of the supply?

- ◯ a   The nature of the current and frequency.
- ◯ b   The earth fault loop impedance external to the installation.
- ⦿ c   Main switch current rating.
- ◯ d   The nominal voltage.

Answer c

Regulation 313.1 shows that options a, b and d are characteristics of the supply, therefore option c must be the answer. Main switch current rating is a feature of a particular item of equipment of the installation, rather than a characteristic of the supply.

313.1 The following characteristics of the supply or supplies, from whatever source, and the normal range of those characteristics where appropriate, shall be determined by calculation, measurement, enquiry or inspection:
(i)   The nominal voltage(s) and its characteristics including harmonic distortion
(ii)   The nature of the current and frequency
(iii)   The prospective short-circuit current at the origin of the installation
(iv)   The earth fault loop impedance of that part of the system external to the installation, Z_e
(v)   The suitability for the requirements of the installation, including the maximum demand
(vi)   The type and rating of the overcurrent protective device(s) acting at the origin of the installation.

12 BS 7671 requires designers to take into account the frequency and quality of maintenance an installation can reasonably be expected to receive when

○ a   assessing staff numbers
○ b   inspecting and testing
○ c   selecting staff
◉ d   specifying or selecting equipment reliability.

Answer d

Regulation 341.1 requires the designer and installer to take into account the likely maintenance of the installation when selecting equipment. Options a and c are not listed as factors for consideration. Inspection and testing **is** mentioned, but inspection and testing are not a designer's function.

341.1 An assessment shall be made of the frequency and quality of maintenance the installation can reasonably be expected to receive during its intended life. The person or body responsible for the operation and/or maintenance of the installation shall be consulted. Those characteristics are to be taken into account in applying the requirements of Parts 4 to 7 so that, having regard to the frequency and quality of maintenance expected:
(i)   any periodic inspection and testing, maintenance and repairs likely to be necessary during the intended life can be readily and safely carried out, and
(ii)   the effectiveness of the protective measures for safety during the intended life shall not diminish, and
(iii)   the reliability of equipment for proper functioning of the installation is appropriate to the intended life.

Section 4

13 Which one of the following is <u>not</u> part of the requirements for fault protection?

- ○ a   Protective earthing.
- ○ b   Protective equipotential bonding.
- ○ c   Automatic disconnection.
- ◉ d   Protection by insulation of live parts.

Answer d

The methods of fault protection for the protective measure of automatic disconnection of supply are found in Regulation 411.3. Protection by basic insulation of live parts (option d) is not listed under 411.3 as it is a method of basic protection.

14 The maximum disconnection time for a TN system with a nominal voltage of 400 V a.c. to Earth is

- ○ a   0.2 second
- ○ b   0.4 second
- ○ c   0.5 second
- ◉ d   5 seconds.

Answer d

The maximum disconnection time for TN systems is given in Regulation 411.3.2.3, which states 5 seconds for distribution circuits and circuits rated over 32 A.

As the question does not give a circuit rating you must read the question carefully to understand that the maximum value is required and that reference should not be made to Table 41.1, which only relates to final circuits rated up to and including 32 A. The fact that the question gives a voltage to earth of 400 V is not relevant to the maximum disconnection time as this would only affect final circuits up to and including 32 A.

411.3.2.3 In a TN system, a disconnection time not exceeding 5 s is permitted for a distribution circuit and for a circuit not covered by Regulation 411.3.2.2.

15 For a TT system, which one of the following conditions should be fulfilled for each circuit protected by an RCD?

- ◉ a $R_A \times I_{\Delta n} \leq 50\ V$
- ○ b $R_A \times I_{\Delta n} \geq 50\ V$
- ○ c $R_A \times I_d \leq 50\ V$
- ○ d $R_A \times I_d \geq 50\ V$

Answer a

You can find the requirements for protection by automatic disconnection of supply in TT systems stated in Regulation 411.5, and Regulation 411.5.3 provides the answer. This requirement ensures that, if the fault current to earth is insufficient to operate the protective device, the difference in potential between the exposed-conductive-part and true earth will not exceed 50 V. The equation below means that R_A multiplied by $I_{\Delta n}$ must be less than or equal to 50 V.

411.5.3   Where an RCD is used for fault protection, the following conditions shall be fulfilled:

(i)   The disconnection time shall be that required by Regulation 411.3.2.2 or 411.3.2.4, and

(ii)   **$R_A \times I_{\Delta n} \leq 50\ V$**

where:

R_A is the sum of the resistances of the earth electrode and the protective conductor connecting it to the exposed-conductive-parts (in ohms).

$I_{\Delta n}$ is the rated residual operating current of the RCD.

16 A 32 A type B circuit-breaker is used to give a disconnection time of 5 seconds in a reduced low voltage system with a nominal voltage to Earth (U_0) of 55 V. What is the maximum value of earth fault loop impedance (Z_s)?

- ○ a   0.44 Ω.
- ◉ b   0.34 Ω.
- ○ c   0.17 Ω.
- ○ d   0.09 Ω.

Answer b

Reference to the index for 'Reduced low voltage system' will direct you to Regulation 411.8 and Table 41.6. See column 2 of Table 41.6, which gives a maximum value for a 32 A type B circuit-breaker of 0.34 Ω.

Table 41.6 Maximum earth fault loop impedance (Z_S) for 5 s disconnection time and U_0 of 55 V (single-phase) and 63.5 V (three-phase)

Notes

	Circuit-breakers to BS EN 60898 and the overcurrent characteristics of RCBOs to BS EN 61009-1 Type						General purpose (gG) fuses to BS 88-2 – fuse systems E and G	
	B		C		D			
U_0 (Volts)	55	63.5	55	63.5	55	63.5	55	63.5
Rating (amperes)				Z_S ohms				
6	1.83	2.12	0.92	1.07	0.47	0.53	3.06	3.53
10	1.10	1.27	0.55	0.64	0.28	0.32	1.72	1.98
16	0.69	0.79	0.34	0.40	0.18	0.20	1.00	1.15
20	0.55	0.64	0.28	0.32	0.14	0.16	0.71	0.81
25	0.44	0.51	0.22	0.26	0.11	0.13	0.55	0.63
32	**0.34**	0.40	0.17	0.20	0.09	0.10	0.44	0.51
40	0.28	0.32	0.14	0.16	0.07	0.08	0.32	0.37
50	0.22	0.25	0.11	0.13	0.06	0.06	0.25	0.29
63	0.17	0.20	0.09	0.10	0.04	0.05	0.20	0.23
80	0.14	0.16	0.07	0.08	0.04	0.04	0.14	0.16
100	0.11	0.13	0.05	0.06	0.03	0.03	0.10	0.12
125	0.09	0.10	0.04	0.05	0.02	0.03	0.08	0.09

From the *IET Wiring Regulations*, Table 41.6, page 61

17 A SELV source can be derived from which one of the following?

- ○ a   Double-wound transformer.
- ○ b   Autotransformer.
- ◉ c   Safety isolating transformer.
- ○ d   Step-up transformer.

Answer c

There are three basic requirements for SELV: (i) the voltage does not exceed extra-low voltage, (ii) the supply is from a safety source and (iii) the live parts are insulated from Earth.

The sources for SELV are specified in Regulation 414.3. Only option c – 'safety isolating transformer' – is listed.

414.3 Sources for SELV and PELV

The following sources may be used for SELV and PELV systems:

(i)   **A safety isolating transformer in accordance with BS EN 61558-2-6**

(ii)   A source of current providing a degree of safety equivalent to that of the safety isolating transformer specified in (i) (eg motor-generator with windings providing equivalent isolation)

(iii)   An electrochemical source (eg a battery) or another source independent of a higher voltage circuit (eg a diesel-driven generator)

(iv)   Certain electronic devices complying with appropriate standards, where provisions have been taken to ensure that, even in the case of an internal fault, the voltage at the outgoing terminals cannot exceed the values specified in Regulation 414.1.1. Higher voltages at the outgoing terminals are, however, permitted if it is ensured that, in case of contact with a live part or in the event of a fault between a live part and an exposed-conductive-part, the voltage at the output terminals is immediately reduced to the value specified by Regulation 414.1.1 or less.
NOTE 1: Examples of such devices include insulation testing equipment and monitoring devices.
NOTE 2: Where higher voltages exist at the outgoing terminals, compliance with this regulation may be assumed if the voltage at the outgoing terminals is within the limits specified in Regulation 414.1.1 when measured with a voltmeter having an internal resistance of at least 3000 ohms.

A mobile source supplied at low voltage, eg a safety isolating transformer or a motor-generator, shall be selected and erected in accordance with the requirements for protection by the use of double or reinforced insulation (see Section 412).

18 Which one of the following <u>cannot</u> be used as basic protection?

- ○ a   Insulation of live parts.
- ○ b   Barriers or enclosures.
- ◉ c   Protective earthing and bonding.
- ○ d   Obstacles.

Answer c
Options a, b and d – insulation of live parts, barriers or enclosures, and obstacles – all provide basic protection (see Sections 416 and 417). Option c, protective earthing and protective bonding, cannot be used as basic protection – these provide fault protection where the protective measure automatic disconnection of supply is employed.

19 In order to provide basic protection, a horizontal top surface of a barrier or enclosure that is readily accessible shall provide a minimum degree of protection of

- ○ a   IPXXA or IP1X
- ○ b   IPXXB or IP2X
- ○ c   IPXXC or IP3X
- ◉ d   IPXXD or IP4X

Answer d
Although BS 7671 does not list all the IP codes, reference to some of them, as appropriate, is made. The most frequently quoted are IPXXB or IP2X. However, in this case the correct answer is IPXXD or IP4X (see Regulation 416.2.2).

416.2.2 A horizontal top surface of a barrier or enclosure which is readily accessible shall provide a degree of protection of at least IPXXD or IP4X.

20 Where arcs, sparks or particles at high temperature may be emitted by fixed equipment in normal service, the equipment shall be

- ◉ a   totally enclosed in arc-resistant material
- ○ b   protected by a 30 mA RCD
- ○ c   enclosed to at least IP55
- ○ d   accessible only by use of a key or tool.

Answer a
This comes under Section 421, Protection against fire caused by electrical equipment, and in particular, Regulation 421.1.3.

Notes

421.1.3 Where arcs, sparks or particles at high temperature may be emitted by fixed equipment in normal service, the equipment shall meet one or more of the following requirements. It shall be:
(i)   totally enclosed in arc-resistant material
(ii)   screened by arc-resistant material from materials upon which the emissions could have harmful effects
(iii)   mounted so as to allow safe extinction of the emissions at a sufficient distance from materials upon which the emissions could have harmful effects
(iv)   in compliance with its standard.

Arc-resistant material used for this protective measure shall be non-ignitable, of low thermal conductivity and of adequate thickness to provide mechanical stability.

21 Except where otherwise recommended by the manufacturer, spotlights and projectors rated at over 100 W and up to 300 W shall be installed at a minimum distance from combustible materials of

○ a   0.5 m
◑ b   0.6 m
◉ c   0.8 m
○ d   1.0 m.

Answer c
See Regulation 422.3.1, item (ii), which shows that option c is correct. Regulation 422.4.2 specifies similar installation requirements.

422.3.1 Except for equipment for which an appropriate product standard specifies requirements, a luminaire shall be kept at an adequate distance from combustible materials. Unless otherwise recommended by the manufacturer, a small spotlight or projector shall be installed at the following minimum distance from combustible materials:
(i)   Rating up to 100 W   0.5 m
(ii)   Over 100 and up to 300 W   0.8 m
(iii)   Over 300 and up to 500 W   1.0 m

22 In locations with increased risks of fire, motors which are automatically or remotely controlled, or which are not continuously supervised, shall be protected against excessive temperature by

- ○ a a protective device that is automatically reset
- ◉ b a protective device with manual reset
- ○ c electronic monitoring equipment that resets
- ○ d electronic monitoring equipment that restarts the motor.

Answer b

Motors which are automatically or remotely controlled are required to be protected against excessive temperature by a protective device with a manual reset (option b), otherwise the overheating will not be identified nor action taken. See Regulation 422.3.7.

422.3.7 A motor which is automatically or remotely controlled or which is not continuously supervised shall be protected against excessive temperature by a protective device with manual reset. A motor with star-delta starting shall be protected against excessive temperature in both the star and delta configurations.

23 Where particular risks of fire exist, the classification for high density occupation areas with easy conditions of evacuation is

- ○ a BD1
- ○ b BD2
- ◉ c BD3
- ○ d BD4.

Answer c

See Regulation 422.2, Conditions for evacuation in an emergency, which gives the precautions where particular risks of fire exist. Note that option a, BD1, is not a relevant code.

422.2 Conditions for evacuation in an emergency

The following regulations refer to conditions:

BD2: Low density occupation, difficult conditions of evacuation

BD3: High density occupation, easy conditions of evacuation

BD4: High density occupation, difficult conditions of evacuation

(Refer to Appendix 5.)

NOTE: Authorities such as those responsible for building construction, public gatherings, fire prevention, hospitals, etc. may specify which BD condition is applicable.

24 When considering protection against overload, the symbol for the current ensuring effective operation of the protective device is

○ a I_b
○ b I_z
○ c I_n
◉ d I_2.

Answer d

Section 433 covers protection against overload current. See Regulation 433.1.1 (iii), which states that I_2 is the symbol for the current ensuring effective operation of the protective device.

25 For protection against overvoltage a 230 V electricity meter should have an impulse withstand of

◉ a 6 kV
○ b 4 kV
○ c 2.5 kV
○ d 1.5 kV.

Answer a

A 230 V electricity meter is a category IV item and as such should have an impulse withstand of 6 kV. See Table 44.4 to identify the category and then Table 44.3 to find the impulse withstand voltage.

26 What is the impulse category of equipment that is part of the fixed electrical installation and other equipment where a high degree of availability is expected?

○ a I.
○ b II.
◉ c III.
○ d IV.

Answer c

Section 443 (Protection Against Overvoltages of Atmospheric Origin or due to Switching) specifies minimum impulse withstands and impulse categories and Table 44.4 provides examples of withstand categories. 'Equipment which is part of the fixed electrical installation and other equipment where a high degree of availability is expected' is an example of category III.

Table 44.4 Examples of various impulse category equipment

Category	Example
I	Equipment intended to be connected to the fixed electrical installation where protection against transient overvoltage is external to the equipment, either in the fixed installation or between the fixed installation and the equipment. Examples of equipment are household appliances, portable tools and similar loads intended to be connected to circuits in which measures have been taken to limit transient overvoltages.
II	Equipment intended to be connected to the fixed electrical installation eg household appliances, portable tools and similar loads, the protective means being either within or external to the equipment.
III	**Equipment which is part of the fixed electrical installation and other equipment where a high degree of availability is expected,** eg distribution boards, circuit-breakers, wiring systems, and equipment for industrial uses, stationary motors with permanent connection to the fixed installation.
IV	Equipment to be used at or in the proximity of the origin of the electrical installation upstream of the main distribution board, eg electricity meter, primary overcurrent device, ripple control unit.

From the *IET Wiring Regulations*, Table 44.4, page 94

---

27 Which of the following is <u>not</u> considered as a potential source of electromagnetic disturbance?

○ a   Electric motors.
○ b   Rectifiers.
◉ c   Electric fires.
○ d   Lifts.

Answer c
The answer is found under 'Measures against electromagnetic disturbances' Chapter 444. The sources of such disturbances are listed in Regulation 444.4.1. Option c is clearly not included as it is not an inductive load.

Notes

Section 5

28 A durable label to BS 951 with the words 'Safety Electrical Connection – Do Not Remove' is <u>not</u> required at

- ○ a   the connection of every earthing conductor to an earth electrode
- ○ b   the point of connection of every bonding conductor to an extraneous-conductive-part
- ○ c   the main earth terminal, where separate from the main switchgear
- ◉ d   a main earthing bar contained within switchgear.

Answer d

You can find the requirements in Regulation group 514.13 (Warning notices: earthing and bonding connections), Regulation 514.13.1. Option d, 'a main earthing bar contained within switchgear', is not included in the list as only skilled persons should gain access, and they will know the importance of maintaining the connections with earth.

514.13.1   A permanent label to BS 951 with the words 'Safety Electrical Connection – Do Not Remove' shall be permanently fixed in a visible position at or near:

- (I)   the point of connection of every earthing conductor to an earth electrode, and
- (ii)   the point of connection of every bonding conductor to an extraneous-conductive-part, and
- (iii)   the main earth terminal, where separate from main switchgear.

29 With regards to temporary overvoltages (TOVs), a surge protective device (SPD) should be selected and erected in accordance with the

- ○ a   ESQCR
- ○ b   DNO
- ○ c   requirements of BS EN 60309
- ◉ d   manufacturer's instructions.

Answer d

This comes under 'Devices for protection against overvoltages' and the answer is clearly shown in Regulation 534.2.3.3, which states that 'SPDs shall be selected and erected in accordance with manufacturers' instructions'. Options a and b relate to the Distribution system and option c is clearly incorrect as BS EN 60309 relates to industrial plugs and sockets.

30 A firefighter's switch shall be provided in the low voltage circuit supplying exterior electrical installations and interior discharge lighting operating at

- ◉ a   a voltage exceeding low voltage
- ○ b   low voltage
- ○ c   voltage band II
- ○ d   medium voltage.

Answer a

A firefighter's switch is required in any low voltage circuit supplying exterior electrical installations operating at a voltage exceeding low voltage (that is, high voltage): see Regulation 537.6.1. This means a firefighter's switch is required for all high voltage lighting (exceeding 1000 V) including neon signs.

537.6.1   A firefighter's switch shall be provided in the low voltage circuit supplying:

(i)   **exterior electrical installations operating at a voltage exceeding low voltage**, and

(ii)   interior discharge lighting installations operating at a voltage exceeding low voltage.

For the purpose of this regulation, an installation in a covered market, arcade or shopping mall is considered to be an exterior installation. A temporary installation in a permanent building used for exhibitions is considered not to be an exterior installation.

31 A plug and socket-outlet may be used for switching off for mechanical maintenance as long as it does not have a rating exceeding

- ○ a   13 A
- ◉ b   16 A
- ○ c   32 A
- ○ d   45 A.

Answer b

The requirements for switching off for mechanical maintenance are given in Section 537. The use of a plug and socket-outlet is allowed up to 16 A. See Regulation 537.3.2.6.

537.3.2.6 A plug and socket-outlet or similar device of rating not exceeding 16 A may be used as a device for switching off for mechanical maintenance.

32 If an area within an installation undergoes a 10 °C rise in ambient temperature, the effect on the current-carrying capacity of cables will be to

- ⦿ a   decrease the value of I_z
- ◯ b   increase the value of I_z
- ◯ c   leave I_z unchanged
- ◯ d   increase the fault current by 10 per cent.

Answer a

I_z is the current-carrying capacity of a cable for continuous service in its particular installation conditions, that is after the relevant rating factors have been applied. Appendix 4 gives guidance on cable ratings and the effect of ambient temperature: see paragraph 2.1, p305, and Tables 4B1 and 4B2. An increase in ambient temperature reduces the ambient temperature rating factor, which reduces I_z. The fault current is affected by the impedance of the cable, which is dependent upon the conductor operating temperature. An increase in the ambient temperature is likely to increase the conductor temperature, increase its impedance and so reduce the fault current.

33 A multicore 70 °C thermoplastic cable with 2.5 mm² conductors supplies a single-phase load of 20 A at 230 V a.c. over a distance of 22 metres. The voltage drop in the cable will be

- ◯ a   6 V
- ◯ b   6.6 V
- ⦿ c   7.92 V
- ◯ d   9.2 V.

Answer c

Refer to Table 4D2B of Appendix 4, p335. The tabulated voltage drop for 2.5 mm² cable is 18 mV/A/m.

To calculate the voltage drop in volts, see Appendix 4, Section 6, Tables of Voltage Drop, third paragraph. The tabulated value of voltage drop (mV/A/m) has to be multiplied by the length of run in metres (L), the design current of the circuit (I_b), and divided by 1000 (to convert to volts).

$$\text{Voltage drop} = \frac{(mV/A/m) \times L \times I_b}{1000} \ V$$

So, voltage drop for the question is

$$\text{Voltage drop} = \frac{(18) \times 22 \times 20}{1000} \; \text{V}$$
$$= 7.92 \; \text{V}$$

34 Where a 3-core cable, with cores coloured brown, black and grey, is used as a switch wire for two-way or intermediate control, the terminations of the conductors shall be identified using

- a   red, blue and yellow tape
- b   black tape only on each core
- ⦿ c   brown tape on the black and grey cores
- d   self-colour tape only.

Answer c

See paragraph 4 of Appendix 7 (Harmonized Cable Core Colours) or Regulation 514.4.4, which refers to Table 51. Table 51 states that line conductors in single-phase circuits shall be coloured brown.

514.4.4 Other conductors

Other conductors shall be identified by colour in accordance with Table 51.

Appendix 7
4 Intermediate and two-way switch wires in a new installation or an addition or alteration to an existing installation

Where a three-core cable with cores coloured brown, black and grey is used as a switch wire, all three conductors being line conductors, the black and grey conductors should be marked brown or L at their terminations.

35 Which one of the following cannot be used as an earth electrode?

- a   Earth plates.
- b   Welded reinforcement of concrete embedded in the earth.
- c   Earth tapes.
- ⦿ d   Gas and water utility pipes.

Answer d

'Gas and water utility pipes' (option d) must not be used as a means of earthing, as they may be replaced by plastic pipes during the lifetime of the installation without the knowledge or understanding of the importance to the user (eg householder). See Regulation 542.2.6.

Notes

542.2.6 A metallic pipe for gases or flammable liquids shall not be used as an earth electrode. The metallic pipe of a water utility supply shall not be used as an earth electrode. Other metallic water supply pipework shall not be used as an earth electrode unless precautions are taken against its removal and it has been considered for such a use.

36 An earthing conductor buried in the ground is protected against corrosion by a sheath, but is not protected against mechanical damage. The minimum size copper conductor that may be installed is

○ a   2.5 mm²
◉ b   16 mm²
○ c   25 mm²
○ d   50 mm².

Answer b
Refer to Regulation 542.3.1 and Table 54.1 for the minimum sizes. Other regulations may require larger sizes to be used. Table 54.1 provides the minimum sizes for protection against corrosion and mechanical damage when buried.

542.3.1 Every earthing conductor shall comply with Section 543 and, where PME conditions apply, shall meet the requirements of Regulation 544.1.1 for the cross-sectional area of a main protective bonding conductor. In addition, where buried in the ground, the earthing conductor shall have a cross-sectional area not less than that stated in Table 54.1. For a tape or strip conductor, the thickness shall be such as to withstand mechanical damage and corrosion.

Table 54.1 Minimum cross-sectional area of a buried earthing conductor

	Protected against mechanical damage	Not protected against mechanical damage
Protected against corrosion by a sheath	2.5 mm² copper 10 mm² steel	**16 mm² copper** 16 mm² coated steel
Not protected against corrosion	25 mm² copper 50 mm² steel	

From the *IET Wiring Regulations*, Table 54.1, page 159

37 Assuming that both the line and protective conductors are of the same material, for a line conductor of 10 mm² if the protective conductor is to be selected, the minimum tabulated cross-sectional area of its associated protective conductor is

○ a 6 mm²
◉ b 10 mm²
○ c 16 mm²
○ d 35 mm².

Answer b
The minimum cross-sectional areas are set out in Section 543, Table 54.7. In this case, the size of the protective conductor is the same as the line conductor.

Table 54.7 Minimum cross-sectional area of protective conductor in relation to the cross-sectional area of associated line conductor

Cross-sectional area of line conductor S	Minimum cross-sectional area of the corresponding protective conductor	
	If the protective conductor is of the same material as the line conductor	If the protective conductor is not of the same material as the line conductor
(mm²)	(mm²)	(mm²)
$S \leq 16$	S	$\dfrac{k1 \times S}{k2}$
$16 < S \leq 35$	16	$\dfrac{k1 \times 16}{k2}$
$S > 35$	$\dfrac{S}{2}$	$\dfrac{k1}{k2} \times \dfrac{S}{2}$

From the *IET Wiring Regulations*, Table 54.7, page 162

38 A radial final circuit feeding socket-outlets supplying several items of data processing equipment has a total protective conductor current in normal service of 18 mA. This circuit must have a high integrity protective conductor

○ a of cross-sectional area less than 1 mm²
◉ b connected as a ring
○ c controlled by an isolator
○ d enclosed in insulated conduit only.

Notes

Answer b

This question relates to earthing requirements for the installation of equipment having high protective conductor currents (see Regulation 543.7). The answer concerns socket-outlet final circuits and is found in Regulation 543.7.2.101 (ii) (a). Options a, c and d are not requirements of the Regulations.

543.7.2.101 For a final circuit with a number of socket-outlets or connection units intended to supply two or more items of equipment, where it is known or reasonably to be expected that the total protective conductor current in normal service will exceed 10 mA, the circuit shall be provided with a high integrity protective conductor connection complying with the requirements of Regulations 543.7.1. The following arrangements of the final circuit are acceptable:

(i)   A ring final circuit with a ring protective conductor. Spurs, if provided, require high integrity protective conductor connections complying with the requirements of Regulation 543.7.1

(ii)   A radial final circuit with a single protective conductor:
         (a) the protective conductor being connected as a ring, or
         (b) a separate protective conductor being provided at the final socket-outlet by connection to the metal conduit or ducting, or
         (c) where two or more similar radial circuits supply socket-outlets in adjacent areas and are fed from the same distribution board, have identical means of short-circuit and overcurrent protection and circuit protective conductors of the same cross-sectional area, then a second protective conductor may be provided at the final socket-outlet on one circuit by connection to the protective conductor of the adjacent circuit.

(iii)   Other circuits complying with the requirements of Regulation 543.7.1.

39 Where an uninterruptible power supply is used it shall comply with

○ a   BS 3036
○ b   BS 88-3
○ c   BS EN 60898 and BS EN 61009
◉ d   BS EN 62040-1 and BS EN 62040-3.

Answer d

Details of uninterruptible power supply sources (UPS) are found in Regulation 560.6.12, which states that 'where an uninterruptible power supply is used, it shall comply with BS EN 62040-1 and BS EN 62040-3'. Options a, b and c are all protective devices.

40 Where an autotransformer is connected to a circuit having a neutral conductor, the common terminal of the winding shall be connected to the

◉ a   neutral conductor
○ b   line conductor
○ c   protective conductor
○ d   bonding conductor.

Answer a

The common terminal of an autotransformer is the terminal that is common to the input and output. Regulation 555.1.1 requires this common terminal to be connected to the neutral so that the neutral on the secondary side of the transformer is at neutral potential.

555.1.1 Where an autotransformer is connected to a circuit having a neutral conductor, the common terminal of the winding shall be connected to the neutral conductor.

Connection of common terminal of autotransformer to the neutral of the supply

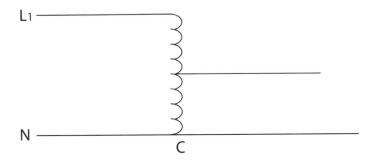

From the *Commentary on BS 7671*, Figure 12.13

41 When selecting wiring systems for safety services, the type of cable that should be used in fire conditions should comply with the test requirements of

○ a   BS 5467
○ b   BS 6231
○ c   IEC 60331-1
◉ d   BS EN 50200.

Answer d

If you refer to Regulation 560.8.1, the only one of the above options listed is d. Option c does not mention test requirements (see Regulation 560.8.1 (iii)).

Notes

560.8.1 One or more of the following wiring systems shall be utilised for safety service required to operate in fire conditions:

(i)   Mineral insulated cable systems complying with BS EN 60702-1 and BS EN 60702-2 and BS EN 60332-1-2

(ii)   Fire-resistant cables complying with IEC 60331-1 or IEC 60331-2 or IEC 60331-3 and with BS EN 60332-1-2

(iii) Fire-resistant cables complying with test requirements of BS EN 50200 or BS 8434 or BS 8491, appropriate for the cable size and with BS EN 60332-1-2

(iv) A wiring system maintaining the necessary fire and mechanical protection.

Section 6

42 When carrying out a visual inspection of an electrical installation, which one of the following does <u>not</u> have to be verified?

- ○ a   The methods of protection against electric shock.
- ◉ b   The electricity supplier.
- ○ c   The connection of conductors.
- ○ d   The presence of undervoltage protective devices.

Answer b

Knowledge of the electricity supplier (option b) is not necessary when carrying out a visual inspection. Regulation 611.3 lists the inspection items, which include options a, c and d.

43 Certain information must be made available to persons carrying out inspection and testing of an installation before the testing commences. One such item of information would be

- ○ a   the name of the client
- ○ b   the name of the person who designed the installation
- ○ c   the length of cable runs in the installation
- ◉ d   any circuit or equipment vulnerable to a typical test.

Answer d

It is necessary to know which circuits and equipment might be vulnerable to the tests of Chapter 61 (option d) before testing commences to avoid causing damage. Refer to Regulations 610.2 and 514.9.1.

Notes

610.2 The result of the assessment of the fundamental principles, Section 131, the general characteristics required by Sections 311 to 313, together with the information required by Regulation 514.9.1, shall be made available to the person or persons carrying out the inspection and testing.

514.9.1 A legible diagram, chart or table or equivalent form of information shall be provided indicating in particular:

(i)   the type and composition of each circuit (points of utilisation served, number and size of conductors, type of wiring), and

(ii)   the method used for compliance with Regulation 410.3.2, and

(iii)   the information necessary for the identification of each device performing the functions of protection, isolation and switching, and its location, and

(iv)   any circuit or equipment vulnerable to a typical test.

---

44 An insulation resistance test is to be carried out on a 3-phase 400 V circuit. The test voltage and minimum acceptable reading would be

○ a   250 V a.c. and 0.5 MΩ

○ b   500 V d.c. and 0.5 MΩ

◉ c   500 V d.c. and 1 MΩ

○ d   800 V d.c. and 0.5 Ω.

Answer c

Insulation resistance is measured with d.c. test voltages to eliminate the effects of capacitance and inductance, so option a is incorrect. Minimum insulation resistance requirements for testing are given in Regulation 612.3.2 and Table 61.

Table 61 Minimum values of insulation resistance

Circuit nominal voltage (V)	Test voltage d.c. (V)	Minimum insulation resistance (MΩ)
SELV and PELV	250	0.5
Up to and including 500 V with the exception of the above systems	**500**	**1.0**
Above 500 V	1000	1.0

From the *IET Wiring Regulations*, Table 61, page 191

Notes

45 The document which must be issued following the periodic inspection and testing of an electrical installation is

- ○ a   a Periodic Inspection Report
- ○ b   a Minor Electrical Installation Works Certificate
- ◉ c   an Electrical Installation Condition Report
- ○ d   an Electrical Installation Certificate.

Answer c

Part 6 Section 634 deals with 'Periodic inspection and testing' and Regulation 634.1 clearly indicates that an Electrical Installation Condition Report is required.

Section 7

46 In a room containing a bath, electrical equipment installed in zone 0 shall have a degree of protection of at least

- ○ a   IPX5
- ○ b   IP5X
- ○ c   IP7X
- ◉ d   IPX7.

Answer d

The requirements for external influences in a bathroom are given in Regulation 701.512.2. For zone 0, the zone which is likely to be totally immersed in water, the requirement is IPX7.

701.512.2 External influences

Installed electrical equipment shall have at least the following degrees of protection:

(i) In zone 0: IPX7

(ii) In zones 1 and 2: IPX4.

47 A flush downlighter in the ceiling less than 2.25 m height above floor level over a bath will need to fulfil the requirements of

- ○ a   zone 2
- ◉ b   zone 1
- ○ c   zone 0
- ○ d   all zones.

Answer b

See Regulation 701.32.1. As the area above a bath is zone 1, then the requirements of zone 1 must apply in this case.

702.32.1 For electrical equipment in parts of walls or ceilings limiting the zones specified in Regulations 701.32.2 to 701.32.4, but being part of the surface of that wall or ceiling, the requirements for the respective zone apply.

48 Which of the following protective measures is permitted in a room containing a bath or shower?

- ○ a   Obstacles.
- ○ b   Placing out of reach.
- ◉ c   Automatic disconnection of supply.
- ○ d   Non-conducting locations.

Answer c

Options a, b and d are not permitted as a protective measure, so option c is correct. See Regulations 701.410.3.5 and 701.410.3.6.

49 In a swimming pool or other basin, the metallic covering or sheath of a wiring system in zones 0, 1 or 2 shall

- ○ a   not be used
- ○ b   not be earthed
- ◉ c   be connected to the supplementary bonding
- ○ d   have reinforced insulation.

Answer c

There are clear requirements for wiring systems in swimming pools in Regulation 702.52 and in particular Regulation 702.522.21.

702.522.21 Erection according to the zones

In zones 0, 1 and 2, any metallic sheath or metallic covering of a wiring system shall be connected to the supplementary equipotential bonding.

NOTE: Cables should preferably be installed in conduits made of insulating material.

50 The requirements of Section 704 of BS 7671 apply to

- ○ a   cloakrooms
- ○ b   offices
- ◉ c   construction and demolition site installations
- ○ d   toilets.

Answer c

There are no additional requirements for the cloakrooms, offices and toilets of construction sites in Section 704. They are not considered part of the building construction site and this is made clear by Regulation 704.1.1. Options a, b and d are excluded by item (vii).

704.1.1 The particular requirements of this section apply to temporary installations for construction and demolition sites during the period of the construction or demolition work, including, for example, the following:

(i)   Construction work of new buildings

(ii)   Repair, alteration, extension, demolition of existing buildings or parts of existing buildings

(iii)   Engineering works

(iv)   Earthworks

(v)   Work of similar nature.

The requirements apply to fixed or movable installations.

The regulations do not apply to:

(vi)   installations covered by the IEC 60621 series 2, where equipment of a similar nature to that used in surface mining applications is involved

(vii) installations in administrative locations of construction sites (eg offices, cloakrooms, meeting rooms, canteens, restaurants, dormitories, toilets), where the general requirements of Part 1 to 6 apply.

51 In locations where livestock is kept, for all circuits other than socket-outlet circuits, an RCD shall be provided with a rating not exceeding

○ a   500 mA

◉ b   300 mA

○ c   100 mA

○ d   30 mA.

Answer b

Section 705, Agricultural and Horticultural Premises, is applicable for locations where livestock is kept. A rated residual operating current $I_{\Delta n}$ not exceeding 300 mA is given in Regulation 705.411.1, item (iii).

705.411.1 General

In circuits, whatever the type of earthing system, the following disconnection device shall be provided:

(i)   In final circuits supplying socket-outlets with rated current not exceeding 32 A, an RCD having the characteristics specified in Regulation 415.1.1

(ii)   In final circuits supplying socket-outlets with rated current more than 32 A, an RCD with a rated residual operating current not exceeding 100 mA

(iii) In all other circuits, RCDs with a rated residual operating current not exceeding 300 mA.

52 Equipment on a pontoon in a marina that is subject to impact to level AG2 should have a mechanical protection code of

- ○ a   IPX4
- ○ b   IPX8
- ○ c   IP55
- ◉ d   IK08.

Answer d

Section 709 gives requirements for marinas and Regulation 709.512.2.1.4 details the requirement for impact protection. See item (iii) which shows option d to be the answer. Options a, b and c are codes for water and foreign solid bodies ingress.

709.512.2.1.4 Impact (AG)

Equipment installed on or above a jetty, wharf, pier or pontoon shall be protected against mechanical damage (Impact of medium severity AG2). Protection shall be afforded by one or more of the following.

(i)   The position or location selected to avoid being damaged by any reasonably foreseeable impact

(ii)   The provision of local or general mechanical protection

(iii)   Installing equipment complying with a minimum degree of protection for external mechanical impact IK08 (see BS EN 62262).

53 Distribution boards for Group 2 medical locations should be clearly labelled and installed

- ○ a   remote from Group 2 locations
- ◉ b   in close proximity to Group 2 locations
- ○ c   within adjacent Group 1 locations
- ○ d   in close proximity to Group 1 locations.

Answer b

The answer will be found in Part 7 'Special installations or locations', Section 710 'Medical locations'. Regulation 710.51.1 shows option b to be the answer.

710.51.1 Distribution boards

Distribution boards shall meet the requirements of BS EN 60439 series.

Distribution boards for Group 2 shall be installed in close proximity to the Group 2 medical locations and clearly labelled.

54 In caravans, each final circuit shall be protected against overcurrent by a device that disconnects

- ⦿ a   all live conductors in that circuit
- ○ b   the line conductors only
- ○ c   line, neutral and protective conductors
- ○ d   the caravan and site supply system.

Answer a

The requirements for electrical installations in caravans are given in Section 721. As the polarity of caravan site supplies is not specified in all countries, caravan vehicle wiring must be suitable for unpolarised supplies, ie any overcurrent devices must disconnect all live conductors. See Regulation 721.43.1.

721.43.1 Final circuits

Each final circuit shall be protected by an overcurrent protective device which disconnects all live conductors of that circuit.

55 Equipment doors or hinged panels in operating or maintenance gangways must be able to open to a minimum of

- ○ a   20 degrees
- ○ b   30 degrees
- ○ c   60 degrees
- ⦿ d   90 degrees.

Answer d

Part 7, Section 729 'Operating and Maintenance Gangways' is referred to for this question. The answer is given in Regulation 729.513.2.

729.513.2 Requirements for operating and maintenance gangways

The width of gangways and access areas shall be adequate for work, operational access, emergency access, emergency evacuation, and for transport of equipment.

Gangways shall permit at least a 90 degree opening of equipment doors or hinged panels (see also Annex A729).

Section 8

56 The requirements of the licensing authority should be ascertained for

- ⊙ a   installations on caravan parks
- ○ b   temporary supplies
- ○ c   types of earthing system to be used
- ○ d   the design of the installation.

Answer a

For installations on caravan parks, the requirements of the licensing authority need to be ascertained before carrying out designs or testing installations. Certificates indicating compliance with BS 7671 may be required. Local authorities may have particular requirements on frequency of testing. Refer to paragraph 6 of Appendix 2 of BS 7671.

Appendix 2

6 For installations in theatres and other places of public entertainment, and on caravan parks, the requirements of the licensing authority should be ascertained. Model Standards were issued by the Department of the Environment in 1977 under the Caravan Sites and Control of Development Act 1960 as guidance for local authorities.

57 I_2 can be greater than 1.45 I_z when the overcurrent device

- ○ a   is providing overload protection
- ○ b   is a circuit-breaker
- ○ c   is rated at over 100 A
- ⊙ d   is providing fault current protection only.

Answer d

I_2 can be greater than 1.45 I_z when the overcurrent device 'is providing fault current protection only' (option d). When overload protection is being provided I_2 must be equal to or less than 1.45 I_z. See Section 3 of Appendix 4 and Regulation 434.5.2.

Appendix 4, Section 3, end paragraph

Where the overcurrent device is intended to afford fault current protection only, I_n can be greater than I_z and I_2 can be greater than 1.45 I_z. The protective device must be selected for compliance with Regulation 434.5.2.

Notes

58 External influences coded BE are classified as

- ⦿ a nature of processed or stored materials
- ○ b conditions of evacuation in an emergency
- ○ c movement of air
- ○ d capability of persons.

Answer a

Appendix 5 (Classification of External Influences), page 387, shows that the BE code classifies nature of processed or stored materials.

59 The maximum permitted voltage drop for a 230 V lighting circuit supplied from a public low voltage system is

- ⦿ a 3%
- ○ b 5%
- ○ c 6%
- ○ d 8%.

Answer a

Appendix 4, Section 6.4, Table 4Ab gives maximum permissible voltage drops

Table 4Ab Voltage drop

	Lighting	Other uses
(i) Low voltage installations supplied directly from a public low voltage distribution system	**3%**	5%
(ii) Low voltage installation supplied from private LV supply	6%	8%

From the *IET Wiring Regulations*, Table 4Ab, page 314

60 **An Electrical Installation Certificate should be made out and signed or otherwise authenticated by**

- ○ a   a responsible person
- ○ b   an authorised person
- ◉ c   a competent person
- ○ d   a qualified person.

Answer c

Appendix 6 gives details of 'Model forms for certification and reporting' and item (i) of the Introduction indicates those who should complete Electrical Installation Certificates.

Introduction

(i) The Electrical Installation Certificate required by Part 6 should be made out and signed or otherwise authenticated by a competent person or persons in respect of the design, construction, inspection and testing of the work.

Answer key

Sample test 1

Question	Answer	Question	Answer
1	d	31	b
2	d	32	a
3	c	33	c
4	a	34	c
5	b	35	d
6	a	36	b
7	c	37	b
8	c	38	b
9	c	39	d
10	a	40	a
11	c	41	d
12	d	42	b
13	d	43	d
14	d	44	c
15	a	45	c
16	b	46	d
17	c	47	b
18	c	48	c
19	d	49	c
20	a	50	c
21	c	51	b
22	b	52	d
23	c	53	b
24	d	54	a
25	a	55	d
26	c	56	a
27	c	57	d
28	d	58	a
29	d	59	a
30	a	60	c

Exam practice 2

Exam practice 2

Sample test 2

The sample test below has 60 questions, the same number as the online exam, and its structure follows that of the online exam. The test appears first without answers, so you can use it as a mock exam. It is then repeated with worked-through answers and extracts from the *IET Wiring Regulations*. Finally, there is an answer key for easy reference.

Answer the questions by filling in the circle next to your chosen option.

Section 1

1   BS 7671 includes the requirements for

○ a   fixed wiring for information and communication technology, signalling and control
○ b   radio interference suppression equipment
○ c   systems for the distribution of electricity to the public
○ d   equipment on board ships and aircraft.

2   BS 7671 requires protection against electric shock to be provided by basic and fault protection. One method common to both is

○ a   the use of RCDs
○ b   the use of Class II equipment
○ c   preventing current from passing through any person or livestock
○ d   the use and implementation of equipotential bonding.

3   Which one of the following types of electrical installation is <u>not</u> covered by BS 7671?

○ a   High protective conductor current installations.
○ b   Lightning protection of buildings.
○ c   Conducting locations with restricted movement.
○ d   Highway power supplies.

4   The selection of the type of wiring and method of installation is <u>not</u> influenced by

○ a   the nature of the location
○ b   the load current
○ c   the value of the prospective short-circuit current
○ d   the nature of the structure supporting the wiring.

Section 2

5 Band II voltage exceeds

- ○ a 2500 V d.c. between live conductors
- ○ b 2500 V d.c. between conductors and Earth
- ○ c 1000 V a.c. rms
- ○ d 120 V d.c. ripple-free.

6 A protective conductor connecting exposed-conductive-parts of equipment to the main earthing terminal is known as the

- ○ a earthing conductor
- ○ b main bonding conductor
- ○ c circuit protective conductor
- ○ d bonding conductor.

Section 3

7 The combination of neutral and protective functions in a single conductor in an earthing system is denoted by the letter

- ○ a S
- ○ b C
- ○ c T
- ○ d N.

8 To ensure continuity of supply to one part of the installation in the event of failure of another part, one approach is

- ○ a for a separate circuit to be installed for each part
- ○ b to use protective equipotential bonding
- ○ c to install larger fuses
- ○ d to install a 100 mA S type RCD.

9 An assessment must be made of the harmful effects of harmonic currents of equipment on other electrical equipment. This assessment is a check of

- ○ a external influences
- ○ b compatibility
- ○ c diversity
- ○ d maintainability.

Notes

10 Every installation must be divided into circuits as necessary in order to

○ a   reduce the cost of installation
○ b   make installation easier
○ c   install a cooker
○ d   facilitate safe inspection, testing and maintenance.

11 Which one of the following is <u>not</u> a potential characteristic of equipment to be considered for harmful effects on other equipment?

○ a   Transient overvoltages.
○ b   d.c. feedback.
○ c   Prospective short-circuit current.
○ d   Excessive protective conductor current.

12 When making an assessment of the frequency and quality of maintenance, a factor to be considered is that

○ a   power factor is monitored
○ b   protective measures for safety remain effective
○ c   starting currents are at a minimum
○ d   unbalanced loads need to be checked more frequently.

Section 4

13 A BS 88-2 Type E 32 A fuse with U_o of 230 V has a maximum tabulated Z_s for 5 seconds disconnection of

○ a $0.36\,\Omega$
○ b $1.04\,\Omega$
○ c $1.44\,\Omega$
○ d $1.84\,\Omega$.

14 Which of the following is a requirement in terms of fault protection where the protective measure of electrical separation is employed?

○ a   The voltage of the separated circuit is not more than 500 V.
○ b   Flexible cables and cords liable to mechanical damage are concealed.
○ c   Live parts are connected to a protective conductor.
○ d   Exposed-conductive-parts are connected to Earth.

15 An RCD can be used as additional protection in the event of failure of the provision for basic protection when the RCD has a rating not exceeding

- ○ a   30 mA
- ○ b   100 mA
- ○ c   300 mA
- ○ d   500 mA.

16 Basic protection can be provided by

- ○ a   protective earthing
- ○ b   barriers
- ○ c   automatic disconnection in case of a fault
- ○ d   functional earth.

17 An obstacle is <u>not</u> intended to

- ○ a   prevent unintentional bodily approach to live parts
- ○ b   prevent intentional contact with live parts during operation
- ○ c   be removed by use of a key or tool
- ○ d   be secured so as to prevent unintentional removal.

18 For a non-conducting location, the resistance of an insulating floor or wall at every point of measurement specified in Part 6 shall not be less than

- ○ a   45 kΩ where the voltage to earth does not exceed 230 V
- ○ b   50 kΩ where the voltage to earth does not exceed 500 V
- ○ c   55 kΩ where the voltage to earth does not exceed 750 V
- ○ d   60 kΩ where the voltage to earth does not exceed 1000 V.

19 Where electrical separation is used to supply more than one item of current-using equipment

- ○ a   warning notices may be omitted
- ○ b   the length of the wiring system should not exceed 500 m
- ○ c   socket-outlets should not have a protective conductor contact
- ○ d   the provisions for basic protection are not required.

20 In locations with risks of fire due to the nature of processed or stored materials, enclosures such as heaters and resistors shall not attain in normal conditions higher surface temperatures than

○ a   90 °C
○ b   85 °C
○ c   60 °C
○ d   55 °C.

21 In a paper mill installation wired in PVC single-core cables in conduit and trunking, where a fire may be caused by a resistive fault in overhead heaters with film elements, protection against an insulation fault shall be provided by a

○ a   BS 88 fuse
○ b   BS EN 60898 cb
○ c   300 mA RCD
○ d   30 mA RCD.

22 The maximum temperature for the exposed metallic casing of a wall-mounted, cord-operated infra-red heater is

○ a   55 °C
○ b   65 °C
○ c   70 °C
○ d   80 °C.

23 Where fault current protection is <u>not</u> required, protection against overload current is generally provided by

○ a   BS EN 60898 fuses
○ b   an inverse-time-lag protective device
○ c   a BS 3036 semi-enclosed fuse
○ d   RCDs.

24 Regulations 434.2 and 434.2.1 do <u>not</u> allow the fault current protective device to be placed on the load side of the reduction in current-carrying capacity of a conductor if

- ○ a   the length of conductor is less than 3 m
- ○ b   the length of conductor exceeds 3 m
- ○ c   it is erected in such a manner as to reduce to a minimum the risk of fault current
- ○ d   it is erected in such a manner as to reduce to a minimum the risk of fire or danger to persons.

25 In the formula

$$t = \frac{k^2 S^2}{I^2}$$

what is the k factor for copper conductors insulated with 70 °C thermoplastic and an assumed initial temperature of 70 °C?

- ○ a   100/86.
- ○ b   115/103.
- ○ c   141.
- ○ d   143.

26 What is the required minimum impulse withstand voltage for category IV equipment for use in an installation with nominal voltage of 400/690?

- ○ a   2.5 kV.
- ○ b   4 kV.
- ○ c   6 kV.
- ○ d   8 kV.

27 Where measures are taken against electromagnetic disturbances, what are the minimum nominal dimensions of a flat cross section bonding ring network conductor?

- ○ a   6 mm x 4 mm.
- ○ b   10 mm x 5 mm.
- ○ c   16 mm x 8 mm.
- ○ d   25 mm x 3 mm.

Section 5

28 Unless otherwise confirmed as suitable, switchgear, protective devices and accessories shall not be connected to conductors intended to operate at a temperature in excess of

- ○ a   30 °C
- ○ b   50 °C
- ○ c   70 °C
- ○ d   90 °C.

29 If the cross-sectional area of a line conductor is 10mm² the minimum size of an associated Surge Protective Device (SPD) connecting conductor is

- ○ a   2.5 mm²
- ○ b   4.0 mm²
- ○ c   6.0 mm²
- ○ d   10.0 mm².

30 A single-phase load of 13 A is supplied via a 16 A BS 88-3 fuse using single-core 70 °C thermoplastic copper cables installed to method 4 (Reference Method B). The rating factor for grouping is 0.7 and for ambient temperature 0.87, and overload protection is to be provided. The minimum acceptable size cable would be

- ○ a   1 mm²
- ○ b   1.5 mm²
- ○ c   2.5 mm²
- ○ d   4 mm².

31 The minimum information contained on a periodic inspection and testing notice would be the

- ○ a   date of inspection and inspector's name
- ○ b   date of next inspection and client's name
- ○ c   date of last inspection and date of next inspection
- ○ d   company's address and date of next inspection.

32 When conductors are identified by numbers, the neutral shall be identified by the number

- ○ a   0
- ○ b   4
- ○ c   5
- ○ d   14.

33 Which one of the following devices would <u>not</u> be suitable to provide overcurrent protection?

○ a   Residual current device to BS 4293.
○ b   Cartridge fuse to BS 1362.
○ c   Cartridge fuse to BS 88-6.
○ d   Rewirable fuse to BS 3036.

34 Under fault conditions, cables are subjected to stress and thermal effects. One item <u>not</u> normally encountered would be

○ a   electrochemical effects
○ b   electromechanical stresses
○ c   electromagnetic effects
○ d   thermal damage.

35 Every connection between conductors, or between conductors and equipment, need <u>not</u> be selected to account for the

○ a   conductor length
○ b   material of the conductor
○ c   number and shape of the wires forming the conductor
○ d   cross-sectional area of the conductor.

36 Where more than one firefighter's switch is installed on any one building, each switch must be

○ a   not more than 3.75 m from the ground
○ b   in a locked location to prevent nuisance operation
○ c   electrically linked
○ d   clearly marked.

37 Where safety depends upon the direction of rotation of a motor, provision shall be made to

○ a   measure the maximum and minimum speed
○ b   monitor the slip frequency
○ c   prevent reverse operation
○ d   prevent operation of overload devices.

Notes

38 A 16 mm² multicore cable incorporating a copper protective conductor is insulated with 90 °C thermosetting insulation. Given a fault current of 500 A and a disconnection time of 0.4 second, the minimum acceptable size of cpc would be

- ○ a   1.5 mm²
- ○ b   2.5 mm²
- ○ c   4.0 mm²
- ○ d   6.0 mm².

39 When using the alternative method to Regulation 543.1.3 of sizing a copper protective conductor, the minimum size to be used when a copper line conductor has a csa of 50 mm² would be

- ○ a   16 mm²
- ○ b   25 mm²
- ○ c   35 mm²
- ○ d   50 mm².

40 The maximum current rating of a protective device protecting a circuit including B22 bayonet cap lampholders would be

- ○ a   5 A
- ○ b   6 A
- ○ c   16 A
- ○ d   20 A.

41 Which of the following is <u>not</u> a BS 7671 requirement when installing a heating cable laid directly in soil?

- ○ a   It is completely embedded in the soil.
- ○ b   It does not suffer damage in the event of normal movement.
- ○ c   It complies with the manufacturer's instructions.
- ○ d   It is protected by a 500 mA RCD.

Section 6

42 During the installation inspection process, which one of the following need <u>not</u> be checked?

- ○ a   The external earth fault loop impedance.
- ○ b   Equipment in compliance with Section 511.
- ○ c   Correct selection and erection.
- ○ d   Visible damage or defects to installed equipment.

43 When testing SELV circuits, the minimum acceptable insulation resistance value is

- ○ a   0.25 MΩ
- ○ b   0.5 MΩ
- ○ c   1 MΩ
- ○ d   2 MΩ.

44 Defects identified during an initial verification should be

- ○ a   noted on the certificate, and the client should be informed
- ○ b   pointed out to the main contractor when the certificate has been completed
- ○ c   made good before the certificate is issued
- ○ d   put on a snagging list to be rectified later.

45 Which of the following is <u>not</u> a requirement for periodic inspection and testing?

- ○ a   Ensuring the safety of persons and livestock from shock and burns.
- ○ b   Ensuring the property is protected against damage from fire and heat due to an installation fault.
- ○ c   Confirmation that the installation is not damaged as to impair safety.
- ○ d   Confirmation that the supply to the installation meets all the requirements of the Electricity Safety, Quality and Continuity Regulations.

Section 7

46 For an electric floor heating system in a bathroom, a fine mesh metallic grid is <u>not</u> required to be connected to the protective conductor of the supply circuit if the protective measure used is

- ○ a   SELV
- ○ b   electrical separation
- ○ c   a non-conducting location
- ○ d   an earth-free equipotential zone.

47 Access to the space under a bath is gained by using a screwdriver to remove the bath panel retaining clips. This space is classified as

- ○ a   zone 0
- ○ b   zone 1
- ○ c   outside the zones
- ○ d   zone 2.

48 For SELV and PELV circuits in a bathroom, basic protection for equipment in all zones shall be provided by

- ○ a obstacles
- ○ b barriers or enclosures
- ○ c non-conducting location
- ○ d placing out of reach.

49 Electrical equipment in zones 0 and 1 of a fountain shall have mechanical protection to a minimum classification of

- ○ a AG0
- ○ b AG1
- ○ c AG2
- ○ d AG3.

50 Which of the following would <u>not</u> be a requirement for an Assembly for Construction Sites (ACS), supplying mobile current-using equipment on a construction site?

- ○ a Main equipotential bonding.
- ○ b Overcurrent protective devices.
- ○ c Devices for fault protection.
- ○ d Socket-outlets.

51 Which one of the following is <u>not</u> included in the scope of Section 705?

- ○ a Chicken-houses.
- ○ b Stables.
- ○ c Piggeries.
- ○ d Staff residences.

52 In conducting locations with restricted movement, supplies to handlamps shall be protected by

- ○ a SELV
- ○ b reduced low voltage supplies
- ○ c RCD or RCBO
- ○ d PELV.

53 The minimum height of overhead conductors on caravan sites, when installed in vehicle movement areas, is

○ a   3.0 m
○ b   3.5 m
○ c   5.8 m
○ d   6.0 m.

54 The particular requirements for mobile or transportable units apply to

○ a   generating sets
○ b   mobile workshops
○ c   mobile machinery to BS EN 60204-1
○ d   traction equipment of electric vehicles.

55 Electrical equipment in a circus installation must have a degree of protection of at least

○ a   IP33
○ b   IP4X
○ c   IP44
○ d   IPX8.

Section 8

56 The ambient air temperature rating factor for 90 °C thermosetting cables operating in an ambient air temperature of 60 °C is

○ a   0.50
○ b   0.56
○ c   0.63
○ d   0.71.

57 A BS 3036 100 A fuse, when carrying 1800 A, has a disconnection time of

○ a   0.2 second
○ b   0.4 second
○ c   4 seconds
○ d   5 seconds.

Notes

58 A 2.5 mm² thermoplastic insulated and sheathed flat cable with protective conductor is laid in a ceiling beneath thermal insulation 80 mm thick in contact with the ceiling board, as shown in the figure below. What is its installed rating?

- ○ a   17 A.
- ○ b   20 A.
- ○ c   21 A.
- ○ d   27 A.

59 The maximum trip time for a general purpose, non-delay 100 mA RCD to BS EN 61008-1 when subjected to a current of 5 times its rating is

- ○ a   40 ms
- ○ b   150 ms
- ○ c   200 ms
- ○ d   300 ms.

60 The installation reference method for a cable installed under a ceiling is

- ○ a   Method A
- ○ b   Method B
- ○ c   Method C
- ○ d   Method D.

Questions and answers

The questions in sample test 2 are repeated below with worked-through answers and extracts from the *IET Wiring Regulations* where appropriate.

Section 1

1   BS 7671 includes the requirements for

◉ a   fixed wiring for information and communication technology, signalling and control
○ b   radio interference suppression equipment
○ c   systems for the distribution of electricity to the public
○ d   equipment on board ships and aircraft.

Answer a

See Part 1, Chapter 11, Scope. The installations covered by BS 7671 are listed in Regulation 110.1.2 and item (v) is 'fixed wiring for information and communication technology, signalling, control and the like' (option a). Exclusions from the scope of BS 7671 (including options b, c and d) can be found in Regulation 110.2.

2   BS 7671 requires protection against electric shock to be provided by basic and fault protection. One method common to both is

○ a   the use of RCDs
○ b   the use of Class II equipment
◉ c   preventing current from passing through any person or livestock
○ d   the use and implementation of equipotential bonding.

Answer c

Refer to Chapter 13 (Fundamental Principles), Regulations 131.2.1 and 131.2.2. Preventing a current from passing through the body of any person or livestock (option c) is a method common to both basic protection and fault protection.

131.2.1 Basic protection (protection against direct contact)

NOTE: For low voltage installations, systems and equipment, 'basic protection' generally corresponds to protection against 'direct contact'.

Persons and livestock shall be protected against dangers that may arise from contact with live parts of the installation.

Notes

This protection can be achieved by one of the following methods:

(i) **Preventing a current from passing through the body of any person or any livestock**

(ii) Limiting the current which can pass through a body to a non-hazardous value.

131.2.2 Fault protection (protection against indirect contact)

NOTE: For low voltage installations, systems and equipment, 'fault protection' generally corresponds to protection against 'indirect contact', mainly with regard to failure of basic insulation.

Persons and livestock shall be protected against dangers that may arise from contact with exposed-conductive-parts during a fault.

This protection can be achieved by one of the following methods:

(i) **Preventing a current resulting from a fault from passing through the body of any person or any livestock**

(ii) Limiting the magnitude of a current resulting from a fault, which can pass through a body, to a non-hazardous value

(iii) Limiting the duration of a current resulting from a fault, which can pass through a body, to a non-hazardous time period.

In connection with fault protection, the application of the method of protective equipotential bonding is one of the important principles for safety.

3 **Which one of the following types of electrical installation is <u>not</u> covered by BS 7671?**

○ a   High protective conductor current installations.

◉ b   Lightning protection of buildings.

○ c   Conducting locations with restricted movement.

○ d   Highway power supplies.

Answer b

The requirements for 'Lightning protection of buildings' (option b) are not covered by BS 7671; they can be found in BS EN 62305. See Regulation 110.2, item (ix).

110.2 Exclusions from scope

The Regulations do not apply to the following installations:

(i) Systems for the distribution of electricity to the public

(ii) Railway traction equipment, rolling stock and signalling equipment

(iii) Equipment of motor vehicles, except those to which the requirements of the Regulations concerning caravans or mobile units are applicable

(iv) Equipment on board ships covered by BS 8450

(v) Equipment of mobile and fixed offshore installations

(vi) Equipment of aircraft

(vii) Those aspects of mines and quarries specifically covered by Statutory Regulations

(viii) Radio interference suppression equipment, except so far as it affects safety of the electrical installation

(ix) Lightning protection systems for buildings and structures covered by BS EN 62305

(x) Those aspects of lift installations covered by relevant parts of BS 5655 and BS EN 81-1

(xi) Electrical equipment of machines covered by BS EN 60204

(xii) Electric fences covered by BS EN 60335-2-76.

4 **The selection of the type of wiring and method of installation is <u>not</u> influenced by**

○ a the nature of the location

◉ b the load current

○ c the value of the prospective short-circuit current

○ d the nature of the structure supporting the wiring.

Answer b

The nature of the location (option a), the value of the prospective short-circuit current (option c) and the nature of the structure supporting the wiring (option d) affect decisions on the type and method of wiring, for the wiring has to be suitable for the location, for fixing to the structure and robust enough to withstand fault currents. Option b, the load current, is not an influence. See Regulation 132.7.

132.7 Type of wiring and method of installation

The choice of the type of wiring system and the method of installation shall include consideration of the following:

(i) The nature of the location

(ii) The nature of the structure supporting the wiring

(iii) Accessibility of wiring to persons and livestock

(iv) Voltage

(v) The electromechanical stresses likely to occur due to short-circuit and earth fault currents

(vi) Electromagnetic interference

(vii) Other external influences (eg mechanical, thermal and those associated with fire) to which the wiring is likely to be exposed during the erection of the electrical installation or in service.

Section 2

5 Band II voltage exceeds

- ○ a   2500 V d.c. between live conductors
- ○ b   2500 V d.c. between conductors and Earth
- ○ c   1000 V a.c. rms
- ◉ d   120 V d.c. ripple-free.

Answer d

Part 2, Definitions, states under 'Voltage bands' that low voltage will normally fall within voltage Band II and that Band II voltages do not exceed 1000 V a.c. rms or 1500 V d.c. Part 2 also defines low voltage as exceeding extra-low voltage but not exceeding 1000 V a.c. or 1500 V d.c. Therefore options a, b and c are not correct and the answer is option d.

Voltage bands
Band I
Band I covers:
- installations where protection against electric shock is provided under certain conditions by the value of voltage;
- installations where the voltage is limited for operational reasons (eg telecommunications, signalling, bell, control and alarm installations).

Extra-low voltage (ELV) will normally fall within voltage Band I.

Band II
Band II contains the voltages for supplies to household and most commercial and industrial installations.
Low voltage (LV) will normally fall within voltage Band II.

NOTE: Band II voltages do not exceed 1000 V a.c. rms or 1500 V d.c.

6 A protective conductor connecting exposed-conductive-parts of equipment to the main earthing terminal is known as the

- ○ a   earthing conductor
- ○ b   main bonding conductor
- ◉ c   circuit protective conductor
- ○ d   bonding conductor.

Answer c

See Part 2 Definitions, Circuit protective conductor.

Circuit protective conductor (cpc). A protective conductor connecting exposed-conductive-parts of equipment to the main earthing terminal.

Earthing and protective conductor terms

Section 3

7   **The combination of neutral and protective functions in a single conductor in an earthing system is denoted by the letter**

○ a   S
◉ b   C
○ c   T
○ d   N.

Answer b
Part 3, Chapter 31, Regulation 312.2 'Types of earthing system' lists the letters and their designations. The letter C denotes 'neutral and protective functions combined in a single conductor (PEN conductor)'.

8   **To ensure continuity of supply to one part of the installation in the event of failure of another part, one approach is**

◉ a   for a separate circuit to be installed for each part
○ b   to use protective equipotential bonding
○ c   to install larger fuses
○ d   to install a 100 mA S type RCD.

Answer a

See Regulation 314.2, which shows that the answer to the question is option a.

314.2 Separate circuits shall be provided for parts of the installation which need to be separately controlled, in such a way that those circuits are not affected by the failure of other circuits, and due account shall be taken of the consequences of the operation of any single protective device.

9 **An assessment must be made of the harmful effects of harmonic currents of equipment on other electrical equipment. This assessment is a check of**

- ○ a   external influences
- ◉ b   compatibility
- ○ c   diversity
- ○ d   maintainability.

Answer b

One item of equipment may detrimentally affect other items, but if an item of equipment can operate satisfactorily with another, it is said to be compatible. See Chapter 33 (Compatibility) and Regulation 331.1.

331 Compatibility of Characteristics

331.1 An assessment shall be made of any characteristics of equipment likely to have harmful effects upon other electrical equipment or other services or likely to impair the supply, for example, for co-ordination with concerned parties eg petrol stations, kiosks and shops within shops. Those characteristics include, for example:

(i)   transient overvoltages
(ii)   undervoltage
(iii)   unbalanced loads
(iv)   rapidly fluctuating loads
(v)   starting currents
(vi)   harmonic currents
(vii)   earth leakage currents
(viii) excessive PE conductor current not due to a fault
(ix)   d.c. feedback
(x)   high-frequency oscillations
(xi)   necessity for additional connections to Earth
(xii)   power factor.

For an external source of energy the distributor shall be consulted regarding any equipment of the installation having a characteristic likely to have significant influence on the supply.

10 Every installation must be divided into circuits as necessary in order to

- ○ a   reduce the cost of installation
- ○ b   make installation easier
- ○ c   install a cooker
- ◉ d   facilitate safe inspection, testing and maintenance.

Answer d

The purpose of BS 7671 is to provide for safety, including safe operation and facilities for inspection and testing. See Regulation 314.1, which shows that even though options a, b and c might be true in particular circumstances and so are logical, they are not the mandatory reason.

314.1   Every installation shall be divided into circuits, as necessary, to
- (i)   avoid danger and minimize inconvenience in the event of a fault
- **(ii)   facilitate safe inspection, testing and maintenance (see also Section 537)**
- (iii)   take account of hazards that may arise from the failure of a single circuit such as a lighting circuit
- (iv)   reduce the possibility of unwanted tripping of RCDs due to excessive protective conductor (PE)currents not due to a fault
- (v)   mitigate the effects of electromagnetic disturbances (see also Chapter 44)
- (vi)   prevent the indirect energizing of a circuit intended to be isolated.

11 Which one of the following is <u>not</u> a potential characteristic of equipment to be considered for harmful effects on other equipment?

- ○ a   Transient overvoltages.
- ○ b   d.c. feedback.
- ◉ c   Prospective short-circuit current.
- ○ d   Excessive protective conductor current.

Answer c

Refer to Regulation 331.1 (shown on the previous page). Options a, b and d may be characteristics of equipment that might have harmful effects on other equipment, while option c is a characteristic of a point in the installation.

12 **When making an assessment of the frequency and quality of maintenance, a factor to be considered is that**

○ a   power factor is monitored
◉ b   protective measures for safety remain effective
○ c   starting currents are at a minimum
○ d   unbalanced loads need to be checked more frequently.

Answer b

Maintenance of the installation must be provided with sufficient frequency and to an appropriate standard so that it remains safe and suitable for continued use. See Chapter 34 (Maintainability), Regulation 341.1.

341.1 An assessment shall be made of the frequency and quality of maintenance the installation can reasonably be expected to receive during its intended life. The person or body responsible for the operation and/or maintenance of the installation shall be consulted. Those characteristics are to be taken into account in applying the requirements of Parts 4 to 7 so that, having regard to the frequency and quality of maintenance expected:

(i)   any periodic inspection and testing, maintenance and repairs likely to be necessary during the intended life can be readily and safely carried out, and

(ii)   **the effectiveness of the protective measures for safety during the intended life shall not diminish, and**

(iii)   the reliability of equipment for proper functioning of the installation is appropriate to the intended life.

Section 4

13 **A BS 88-2 Type E 32 A fuse with U_0 of 230 V has a maximum tabulated Z_s for 5 seconds disconnection of**

○ a $0.36\,\Omega$
○ b $1.04\,\Omega$
○ c $1.44\,\Omega$
◉ d $1.84\,\Omega$.

Answer d

You need to look at Table 41.4, which gives the maximum earth fault loop impedances for this nominal line voltage for this type of fuse for a 5 second disconnection time.

Table 41.4 Maximum earth fault loop impedance (Z_s) for fuses, for 5 s disconnection time with U_0 of 230 V (see Regulation 411.4.8)

(a) General purpose (gG) and motor applications (gM) fuses to BS 88-2 – fuse systems E (bolted) and G (clip in)

Rating (amperes)	6	10	16	20	25	32	40	50
Z_s (ohms)	12.8	7.19	4.18	2.95	2.30	**1.84**	1.35	1.04

From the *IET Wiring Regulations*, extract from Table 41.4, page 56

14 Which of the following is a requirement in terms of fault protection where the protective measure of electrical separation is employed?

◉ a   The voltage of the separated circuit is not more than 500 V.
○ b   Flexible cables and cords liable to mechanical damage are concealed.
○ c   Live parts are connected to a protective conductor.
○ d   Exposed-conductive-parts are connected to Earth.

Answer a
Regulation 413.3.2 states that the voltage shall not exceed 500 V.
Options b, c and d contravene the requirements of Regulations 413.3.3, 413.3.4 and 413.3.6.

413.3 Requirements for fault protection
413.3.2 The separated circuit shall be supplied through a source with at least simple separation, and the voltage of the separated circuit shall not exceed 500 V.

15 An RCD can be used as additional protection in the event of failure of the provision for basic protection when the RCD has a rating not exceeding

◉ a   30 mA
○ b   100 mA
○ c   300 mA
○ d   500 mA.

Answer a
See Regulation 415.1.1, which shows that option a is the answer.

415.1.1 The use of RCDs with a rated residual operating current ($I_{\Delta n}$) not exceeding 30 mA and an operating time not exceeding 40 ms at a residual current of 5 $I_{\Delta n}$ is recognised in a.c. systems as additional protection in the event of failure of the provision for basic protection and/or the provision for fault protection or carelessness by users.

Notes

16 Basic protection can be provided by

- ○ a protective earthing
- ◉ b barriers
- ○ c automatic disconnection in case of a fault
- ○ d functional earth.

Answer b

Basic protection is the term given to those protective measures which prevent persons or livestock from coming into direct contact with live parts. Barriers can provide basic protection. See Regulations 416.2.1 to 416.2.5. Options a and c are components of fault protection and option d, 'functional earth', is not associated with safety.

17 An obstacle is <u>not</u> intended to

- ○ a prevent unintentional bodily approach to live parts
- ◉ b prevent intentional contact with live parts during operation
- ○ c be removable by use of a key or tool
- ○ d be secured so as to prevent unintentional removal.

Answer b

This is the subject of basic protection by the rule of obstacles. See Regulation 417.2 which shows that option b is correct.

417.2 Obstacles

NOTE: Obstacles are intended to prevent unintentional contact with live parts but not intentional contact by deliberate circumvention of the obstacle.

417.2.1 Obstacles shall prevent:
(i) unintentional bodily approach to live parts, and
(ii) unintentional contact with live parts during the operation of live equipment in normal service.

417.2.2 An obstacle may be removed without the use of a key or tool but shall be secured so as to prevent unintentional removal.

18 For a non-conducting location, the resistance of an insulating floor or wall at every point of measurement specified in Part 6 shall not be less than

- ○ a 45 kΩ where the voltage to earth does not exceed 230 V
- ◉ b 50 kΩ where the voltage to earth does not exceed 500 V
- ○ c 55 kΩ where the voltage to earth does not exceed 750 V
- ○ d 60 kΩ where the voltage to earth does not exceed 1000 V.

Answer b
Limits are set for the resistance of insulating floors or walls by Regulation 418.1.5.

418.1.5 The resistance of insulating floors and walls at every point of measurement under the conditions specified in Part 6 shall be not less than:

(i) **50 kΩ, where the nominal voltage of the installation does not exceed 500 V**, or

(ii) 100 kΩ, where the nominal voltage of the installation exceeds 500 V.

NOTE: If at any point the resistance is less than the specified value, the floors and walls are deemed to be extraneous-conductive-parts for the purposes of protection against electric shock.

19 Where electrical separation is used to supply more than one item of current-using equipment

○ a   warning notices may be omitted
◉ b   the length of the wiring system should not exceed 500 m
○ c   socket-outlets should not have a protective conductor contact
○ d   the provisions for basic protection are not required.

Answer b
See Regulation 418.3.8, which states that the length of the wiring system shall not exceed 500 m. Also see Regulations 418.3, 418.3.1 and 418.3.5 which show that options a, c and d are incorrect.

20 In locations with risks of fire due to the nature of processed or stored materials, enclosures such as heaters and resistors shall not attain in normal conditions higher surface temperatures than

◉ a   90 °C
○ b   85 °C
○ c   60 °C
○ d   55 °C.

Answer a
In normal locations, BS 7671 accepts the temperature limits set by the British Standard for the particular item of equipment. However, in locations with risks of fire due to the nature of processed or stored materials, BS 7671 sets limits that override equipment standards. See Regulation 422.3.2.

Notes

422.3.2 Measures shall be taken to prevent an enclosure of electrical equipment such as a heater or resistor from exceeding the following temperatures:
(i)   90 °C under normal conditions, and
(ii)   115 °C under fault conditions.

21 In a paper mill installation wired in PVC single-core cables in conduit and trunking, where a fire may be caused by a resistive fault in overhead heaters with film elements, protection against an insulation fault shall be provided by a

○ a   BS 88 fuse
○ b   BS EN 60898 cb
○ c   300 mA RCD
◉ d   30 mA RCD.

Answer d
See Regulation 422.3.9, second paragraph of (i).

422.3.9 Wiring systems, other than mineral insulated cables, busbar trunking systems or powertrack systems shall be protected against insulation faults:
(i)   in a TN or a TT system, by an RCD having a rated residual operating current ($I_{\Delta n}$) not exceeding 300 mA according with Regulation 531.2.4 and to relevant product standards.
       Where a resistive fault may cause a fire, eg for overhead heating with heating film elements, the rated residual operating current shall not exceed 30 mA.
(ii)   in an IT system, by an insulation monitoring device with audible and visual signals provided in accordance with Regulation 538.1. Disconnection times in the event of a second fault are given in Chapter 41. Alternatively, RCDs with a rated residual operating current as specified in (i) may be used. In the event of a second fault see Chapter 41 for disconnection times.

22 The maximum temperature for the exposed metallic casing of a wall-mounted, cord-operated infra-red heater is

○ a   55 °C
○ b   65 °C
○ c   70 °C
◉ d   80 °C.

Answer d

See Section 423, Protection against burns, Table 42.1.

Table 42.1 Temperature limit under normal load conditions for an accessible part of equipment within arm's reach

Accessible part	Material of accessible surfaces	Maximum temperature (°C)
A hand-held part	Metallic Non-metallic	55 65
A part intended to be touched but not hand-held	Metallic Non-metallic	70 80
A part which need not be touched for normal operation	**Metallic** Non-metallic	**80** 90

From the *IET Wiring Regulations*, Table 42.1, page 77

23 Where fault current protection is <u>not</u> required, protection against overload current is generally provided by

- ○ a   BS EN 60898 fuses
- ◉ b   an inverse-time-lag protective device
- ○ c   a BS 3036 semi-enclosed fuse
- ○ d   RCDs.

Answer b

Options a and c will provide both overload and fault protection and option d provides protection against earth faults. See Regulation 432.2.

432.2 Protection against overload current only
A device providing protection against overload current is generally an inverse-time-lag protective device whose rated short-circuit breaking capacity may be below the value of the maximum prospective fault current at the point where the device is installed. Such a device shall satisfy the relevant requirements of Section 433.

24 Regulations 434.2 and 434.2.1 do <u>not</u> allow the fault current protective device to be placed on the load side of the reduction in current-carrying capacity of a conductor if

- ○ a   the length of conductor is less than 3 m
- ◉ b   the length of conductor exceeds 3 m
- ○ c   it is erected in such a manner as to reduce to a minimum the risk of fault current
- ○ d   it is erected in such a manner as to reduce to a minimum the risk of fire or danger to persons.

Answer b

BS 7671 allows the fault current protective device to be placed on the load side of the reduction in current-carrying capacity of a conductor as is often necessary in short cable lengths between busbars and the overcurrent device, providing precautions are taken, such as ensuring that the cable length is no more than 3 metres. See Section 434 and Regulations 434.2 and 434.2.1.

434.2 Position of devices for protection against fault current

A device providing protection against fault current shall be installed at the point where a reduction in the cross-sectional area or other change causes a reduction in the current-carrying capacity of the conductors, except where Regulation 434.2.1, 434.2.2 or 434.3 applies.

434.2.1 Except where Regulation 434.2.2 or 434.3 applies, a device for protection against fault current may be installed other than as specified in Regulation 434.2, under the following conditions:

In the part of the conductor between the point of reduction of cross-sectional area or other change and the position of the protective device there shall be no branch circuits or socket-outlets and that part of the conductor shall:

- **(i)**   **not exceed 3 m in length**, and
- (ii)   be installed in such a manner as to reduce the risk of fault to a minimum, and
  NOTE: This condition may be obtained, for example, by reinforcing the protection of the wiring against external influences.
- (iii)   be installed in such a manner as to reduce to a minimum the risk of fire or danger to persons.

25 In the formula

$$t = \frac{k^2 S^2}{I^2}$$

what is the k factor for copper conductors insulated with 70 °C thermoplastic and an assumed initial temperature of 70 °C?

- ○ a   100/86.
- ◉ b   115/103.
- ○ c   141.
- ○ d   143.

Answer b

Refer to Regulation 434.5.2 and Table 43.1. Be careful that you read off the value for 70 °C thermoplastic.

Table 43.1 Values of k for common materials, for calculation of the effects of fault current for disconnection times up to 5 seconds

Conductor insulation	Thermoplastic				Thermosetting		Mineral insulated	
	90 °C		70 °C		90 °C	60 °C	Thermoplastic sheath	Bare (unsheathed)
Conductor cross-sectional area	≤ 300 mm²	> 300 mm²	≤ 300 mm²	> 300 mm²				
Initial temperature	90 °C		70 °C		90 °C	60 °C	70 °C	105 °C
Final temperature	160 °C	140 °C	160 °C	140 °C	250 °C	200 °C	160 °C	250 °C
Copper conductor	k = 100	k = 86	**k = 115**	**k = 103**	k = 143	k = 141	k = 115	k = 135/115[a]
Aluminium conductor	k = 66	k = 57	k = 76	k = 68	k = 94	k = 93		
Tin soldered joints in copper conductors	k = 100	k = 86	k = 115	k = 103	k = 100	k = 122		

[a] This value shall be used for bare cables exposed to touch.

From the *IET Wiring Regulations*, Table 43.1, page 85

Notes

26 What is the required minimum impulse withstand voltage for category IV equipment for use in an installation with nominal voltage of 400/690?

- ○ a   2.5 kV.
- ○ b   4 kV.
- ○ c   6 kV.
- ● d   8 kV.

Answer d

You need to refer to Section 443 and Table 44.3 for minimum impulse withstand voltage.

Table 44.3   Required minimum impulse withstand voltage

Nominal voltage of the installation V	Required minimum impulse withstand voltage kV [1]			
	Category IV (equipment with very high impulse voltage)	Category III (equipment with high impulse voltage)	Category II (equipment with normal impulse voltage)	Category I (equipment with reduced impulse voltage)
230/240 277/480	6	4	2.5	1.5
400/690	**8**	6	4	2.5
1000	12	8	6	4

From the *IET Wiring Regulations*, Table 44.3, page 94

[1] This impulse withstand voltage is applied between live conductors and PE.

27 Where measures are taken against electromagnetic disturbances, what are the minimum nominal dimensions of a flat cross section bonding ring network conductor?

- ○ a   6 mm x 4 mm.
- ○ b   10 mm x 5 mm.
- ○ c   16 mm x 8 mm.
- ● d   25 mm x 3 mm.

Answer d

As the question suggests this is found under 'measures against electromagnetic disturbances'. The nominal dimensions are given in Regulation 444.5.3.

444.5.3 Sizing and installation of copper bonding ring network conductors

Equipotential bonding designed as a bonding ring network shall have the following minimal nominal dimensions:

(a) Flat cross-section: 25 mm x 3 mm

(b) Round diameter: 8 mm

Bare conductors shall be protected against corrosion at their supports and on their passage through walls.

Section 5

28 Unless otherwise confirmed as suitable, switchgear, protective devices and accessories shall not be connected to conductors intended to operate at a temperature in excess of

- a   30 °C
- b   50 °C
- c   70 °C
- d   90 °C.

Answer c

While thermosetting cables with a conductor operating temperature of 90 °C are commonly available, most equipment is only designed for cables operating at a maximum of 70 °C (option c). Before using a cable loaded so that its conductor temperature exceeds 70 °C, it must be checked with the supplier that the switchgear etc is suitable for the temperature. It may be necessary to increase the cable size to reduce the conductor temperature. Refer to Regulation 512.1.5.

512.1.5 Compatibility

Every item of equipment shall be selected and erected so that it will neither cause harmful effects to other equipment nor impair the supply during normal service including switching operations.

Switchgear, protective devices, accessories and other types of equipment shall not be connected to conductors intended to operate at a temperature exceeding 70 °C at the equipment in normal service unless the equipment manufacturer has confirmed that the equipment is suitable for such conditions, or the conductor size shall be chosen based on the current ratings for 70 °C cables of a similar construction. See also Regulation 523.1 and Table 4A3.

Notes

29 **If the cross-sectional area of a line conductor is 10mm² the minimum size of an associated Surge Protective Device (SPD) connecting conductor is**

- a   2.5 mm²
- ◉ b   4.0 mm²
- c   6.0 mm²
- d   10.0 mm².

Answer b

This refers to devices for protection against overvoltages section 534. The answer is found in Regulation 534.2.10 (i), which requires the Surge Protective Device (SPD) connecting conductor to be not less than 4.0 mm² as the line conductor is 10 mm².

534.2.10   Cross-sectional area of connecting conductors

The connecting conductors of SPDs shall either:

(i)   **have a cross-sectional area of not less than 4.0mm² copper (or equivalent) if the cross-sectional area of the line conductors is greater than or equal to 4.0mm², or**

(ii)   have a cross-sectional area of not less than that of the line conductors, where the line conductors have a cross-sectional area less than 4.0mm²

Where there is a structural lightning protection system, the minimum cross-sectional area for Type 1 SPDs shall be 16mm² copper, or equivalent.

30 **A single-phase load of 13 A is supplied via a 16 A BS 88-3 fuse using single-core 70 °C thermoplastic copper cables installed to method 4 (Reference Method B). The rating factor for grouping is 0.7 and for ambient temperature 0.87, and overload protection is to be provided. The minimum acceptable size cable would be**

- a   1 mm²
- b   1.5 mm²
- c   2.5 mm²
- ◉ d   4 mm².

Answer d

Refer to Appendix 4, 5.1.2 for the calculation.

The procedure is:
(1) the design current I_b of the circuit must be established (I_b is 13 A)
(2) the overcurrent device rating I_n is then selected so that I_n is greater than or equal to I_b

From the question, the device rating I_n is 16 A, which is greater than the load I_b which is 13 A, so this is satisfactory.

The required tabulated current-carrying capacity of the cable is given by $I_t \geq \dfrac{I_n}{C_a C_i C_g C_f}$

where:
C_a is the rating factor for ambient temperature
C_i is the rating factor for thermal insulation
C_g is the rating factor for grouping
C_f is the rating factor for semi-enclosed fuses

In this example C_a is 0.87, and C_g is 0.7. C_i can be assumed to be 1 as no reference to thermal insulation is made and C_f is 1 as rewirable fuses are not used.

Hence $I_t = \dfrac{16}{0.87 \times 0.7} = 26.27$ A

From Table 4D1A of BS 7671, column 4, the cable ratings are:

1 mm²	13.5 A
1.5 mm²	17.5 A
2.5 mm²	24 A
4 mm²	32 A

The answer must be option d.

31 The minimum information contained on a periodic inspection and testing notice would be the

- ○ a date of inspection and inspector's name
- ○ b date of next inspection and client's name
- ◉ c date of last inspection and date of next inspection
- ○ d company's address and date of next inspection.

Answer c

The information required is stated in Regulation 514.12.1. This does not require the inspector's name, client's name or company's address but does require information on the date of last inspection and date of next inspection (option c).

514.12.1 A notice of such durable material as to be likely to remain easily legible throughout the life of the installation, shall be fixed in a prominent position at or near the origin of every installation upon completion of the work carried out in accordance with Chapter 61 or 62. The notice shall be inscribed in indelible characters not smaller than those illustrated here and shall read as follows:

IMPORTANT

This installation should be periodically inspected and tested and a report on its condition obtained, as prescribed in the IET Wiring Regulations BS 7671 Requirements for Electrical Installations.

Date of last inspection ....

Recommended date of next inspection ....

32 When conductors are identified by numbers, the neutral shall be identified by the number

- ◉ a 0
- ○ b 4
- ○ c 5
- ○ d 14.

Answer a

You need to refer to Regulation 514.5.4.

514.5.4 Numeric

Conductors may be identified by numbers, the number 0 being reserved for the neutral or midpoint conductor.

33 **Which one of the following devices would <u>not</u> be suitable to provide overcurrent protection?**

- ⊙ a   Residual current device to BS 4293.
- ○ b   Cartridge fuse to BS 1362.
- ○ c   Cartridge fuse to BS 88-2.
- ○ d   Rewirable fuse to BS 3036.

Answer a

Refer to Part 2 Definitions, which gives the definitions of fuses and residual current devices. Fuses provide overcurrent protection, as the fuse element melts to limit the current. RCDs do not provide overcurrent protection. A residual current device only operates when the algebraic sum of the neutral and line currents is not zero; it will not react to an overcurrent.

Fuse. A device which, by the melting of one or more of its specially designed and proportioned components, opens the circuit in which it is inserted by breaking the current when this exceeds a given value for a sufficient time. The fuse comprises all the parts that form the complete device.

Residual current device (RCD). A mechanical switching device or association of devices intended to cause the opening of the contacts when the residual current attains a given value under specified conditions.

34 **Under fault conditions, cables are subjected to stress and thermal effects. One item <u>not</u> normally encountered would be**

- ⊙ a   electrochemical effects
- ○ b   electromechanical stresses
- ○ c   electromagnetic effects
- ○ d   thermal damage.

Answer a

BS 7671 contains specific requirements with regard to electromechanical stresses, electromagnetic effects and thermal damage (options b, c and d) likely to occur under fault conditions. Option a, electrochemical effects, such as electrolytic corrosion, are only mentioned in relation to the selection and erection of wiring systems and earth arrangements. See Regulations 521.5.1, 521.5.2 and 521.5.100.

Notes

35 Every connection between conductors, or between conductors and equipment, need <u>not</u> be selected to account for the

- ⦿ a conductor length
- ○ b material of the conductor
- ○ c number and shape of the wires forming the conductor
- ○ d cross-sectional area of the conductor.

Answer a

In Section 526, Regulation 526.2 lists a number of considerations, of which options b, c and d are three. Option a is not mentioned as it has no bearing on the suitability of a connection for use.

526.2 The selection of the means of connection shall take account of, as appropriate:

- **(i)**   **the material of the conductor and its insulation**
- **(ii)**   **the number and shape of the wires forming the conductor**
- **(iii)** **the cross-sectional area of the conductor**
- (iv)   the number of conductors to be connected together
- (v)   the temperature attained at the terminals in normal service such that the effectiveness of the insulation of the conductors connected to them is not impaired
- (vi)   the provision of adequate locking arrangements in situations subject to vibration or thermal cycling.

Where a soldered connection is used the design shall take account of creep, mechanical stress and temperature rise under fault conditions.

36 Where more than one firefighter's switch is installed on any one building, each switch must be

- ○ a not more than 3.75 m from the ground
- ○ b in a locked location to prevent nuisance operation
- ○ c electrically linked
- ⦿ d clearly marked.

Answer d

The requirements for firefighter's switches are in Regulation group 537.6. Where there is more than one, each switch must be clearly marked to indicate what it will switch. See Regulation 537.6.3, item (iv).

537.6.3 Every firefighter's switch provided for compliance with Regulation 537.6.1 shall comply with all the relevant requirements of the following items (i) to (iv) and any requirements of the local fire authority:

(i) For an exterior installation, the switch shall be outside the building and adjacent to the equipment, or alternatively a notice indicating the position of the switch shall be placed adjacent to the equipment and a notice shall be fixed near the switch so as to render it clearly distinguishable

(ii) For an interior installation, the switch shall be in the main entrance to the building or in another position to be agreed with the local fire authority

(iii) The switch shall be placed in a conspicuous position, reasonably accessible to firefighters and, except where otherwise agreed with the local fire authority, at not more than 2.75 m from the ground or the standing beneath the switch

(iv) Where more than one switch is installed on any one building, each switch shall be clearly marked to indicate the installation or part of the installation which it controls.

37 **Where safety depends upon the direction of rotation of a motor, provision shall be made to**

○ a   measure the maximum and minimum speed
○ b   monitor the slip frequency
◉ c   prevent reverse operation
○ d   prevent operation of overload devices.

Answer c
Looking up 'Motors – direction of rotation' in the index directs you to Regulation 537.5.4.3, which defines option c as the answer.

537.5.4.3 Where safety depends on the direction of rotation of a motor, provision shall be made for the prevention of reverse operation due to, for example, a reversal of phases.

Notes

38 A 16 mm² multicore cable incorporating a copper protective conductor is insulated with 90 °C thermosetting insulation. Given a fault current of 500 A and a disconnection time of 0.4 second, the minimum acceptable size of cpc would be

- ○ a 1.5 mm²
- ◉ b 2.5 mm²
- ○ c 4.0 mm²
- ○ d 6.0 mm².

Answer b

See Regulation 543.1.3. The fault current I is given in the question, as is the disconnection time t. k must be looked up in Table 54.3. From Table 54.3, the k value for a copper protective conductor incorporated in a cable with 90 °C thermosetting insulation is 143.

Hence:

$$S = \frac{\sqrt{500^2 \times 0.4}}{143} = 2.211 \text{ mm}^2$$

The last sentence of Regulation 543.1.3 states: Where the application of the formula produces a non-standard size, a conductor having the nearest larger standard cross-sectional area shall be used.

This is 2.5 mm².

543.1.3 The cross-sectional area, where calculated, shall be not less than the value determined by the following formula or shall be obtained by reference to BS 7454:

$$S = \frac{\sqrt{I^2 t}}{k}$$

Table 54.3 Values of k for protective conductor incorporated in a cable or bunched with cables, where the assumed initial temperature is 70 °C or greater

Notes

Material of conductor	Insulation material		
	70 °C thermoplastic	90 °C thermoplastic	**90 °C thermosetting**
Copper	115/103*	100/86*	**143**
Aluminium	76/68*	66/57*	94
Assumed initial temperature	70 °C	90 °C	90 °C
Final temperature	160 °C/140 °C*	160 °C/140 °C*	250 °C

* Above 300 mm²

From the *IET Wiring Regulations*, Table 54.3, page 161

39 When using the alternative method to Regulation 543.1.3 of sizing a copper protective conductor, the minimum size to be used when a copper line conductor has a csa of 50 mm² would be

○ a   16 mm²
◉ b   25 mm²
○ c   35 mm²
○ d   50 mm².

Answer b
See Regulation group 543.1 (Cross-sectional areas) for advice on how to determine protective conductor sizes. Regulation 543.1.1 offers two approaches. The first method is calculation using the formula. The alternative method is selection in accordance with Regulation 543.1.4 and Table 54.7.

In the question, S (at 50 mm²) > 35, so from Table 54.7 the minimum protective conductor size is S/2, that is 50/2 or 25mm², therefore option b is the answer.

Notes

543.1.1 The cross-sectional area of every protective conductor, other than a protective bonding conductor, shall be:
(i)   calculated in accordance with Regulation 543.1.3, or
(ii)   selected in accordance with Regulation 543.1.4.

Calculation in accordance with Regulation 543.1.3 is necessary if the choice of cross-sectional area of line conductors has been determined by considerations of short-circuit current and if the earth fault current is expected to be less than the short-circuit current.

543.1.4 Where it is desired not to calculate the minimum cross-sectional area of a protective conductor in accordance with Regulation 543.1.3, the cross-sectional area may be determined in accordance with Table 54.7.

Where the application of Table 54.7 produces a non-standard size, a conductor having the nearest larger standard cross-sectional area shall be used.

Table 54.7   Minimum cross-sectional area of protective conductor in relation to the cross-sectional area of associated line conductor

Cross-sectional area of line conductor S	Minimum cross-sectional area of the corresponding protective conductor	
	If the protective conductor is of the same material as the line conductor	If the protective conductor is not of the same material as the line conductor
(mm²)	(mm²)	(mm²)
$S \leq 16$	S	$\dfrac{k1}{k2} \times S$
$16 < S \leq 35$	16	$\dfrac{k1 \times 16}{k2}$
$S > 35$	$\dfrac{S}{2}$	$\dfrac{k1}{k2} \times \dfrac{S}{2}$

From the *IET Wiring Regulations*,
Table 54.7, page 162

40 The maximum current rating of a protective device protecting a circuit including B22 bayonet cap lampholders would be

- ○ a   5 A
- ○ b   6 A
- ◉ c   16 A
- ○ d   20 A.

Answer c

On the failure of a lamp, a short circuit can develop, which may cause an electric arc that needs to be interrupted by the overcurrent device (before significant overheating of the luminaire occurs). Consequently, a maximum device rating is set by Regulation 559.6.1.6.

559.6.1.6 Lighting circuits incorporating B15, B22, E14, E27 or E40 lampholders shall be protected by an overcurrent protective device of maximum rating 16 A.

41 Which of the following is <u>not</u> a BS 7671 requirement when installing a heating cable laid directly in soil?

- ○ a   It is completely embedded in the soil.
- ○ b   It does not suffer damage in the event of normal movement.
- ○ c   It complies with the manufacturer's instructions.
- ◉ d   It is protected by a 500 mA RCD.

Answer d

This is typical of under-soil heating in a horticultural location for soil warming for plant and seed propagation. See Regulation 554.4.3, where options a, b and c are all stated as requirements.

554.4.3 A heating cable laid directly in soil, a road or the structure of a building shall be installed so that it:
- (i)   is completely embedded in the substance it is intended to heat, and
- (ii)   does not suffer damage in the event of movement normally to be expected in it or the substance in which it is embedded, and
- (iii)   complies in all respects with the manufacturer's instructions and recommendations.

Notes

Section 6

42 During the installation inspection process, which one of the following need not be checked?

- ◉ a   The external earth fault loop impedance.
- ○ b   Equipment in compliance with Section 511.
- ○ c   Correct selection and erection.
- ○ d   Visible damage or defects to installed equipment.

Answer a
Earth fault loop impedance is an instrument test, and is not part of the inspection process.

43 When testing SELV circuits, the minimum acceptable insulation resistance value is

- ○ a   0.25 MΩ
- ◉ b   0.5 MΩ
- ○ c   1 MΩ
- ○ d   2 MΩ.

Answer b
You can find the requirements for minimum values of insulation resistance in Table 61.

612.3.2   The insulation resistance measured with the test voltages indicated in Table 61 shall be considered satisfactory if the main switchboard and each distribution circuit tested separately, with all its final circuits connected but with current-using equipment disconnected, has an insulation resistance not less than the appropriate value given in Table 61.

NOTE: More stringent requirements are applicable for the wiring of fire alarm systems in buildings, see BS 5839-1.

Table 61   Minimum values of insulation resistance

Circuit nominal voltage (V)	Test voltage d.c. (V)	Minimum insulation resistance (MΩ)
SELV and PELV	250	**0.5**
Up to and including 500 V with the exception of the above systems	500	1.0
Above 500 V	1000	1.0

From the *IET Wiring Regulations*, Table 61, page 191

44 Defects identified during an initial verification should be

○ a   noted on the certificate, and the client should be informed
○ b   pointed out to the main contractor when the certificate has been completed
◉ c   made good before the certificate is issued
○ d   put on a snagging list to be rectified later.

Answer c

The general requirements for initial testing are given in Regulation 612.1. See the fifth paragraph. Note that the fault must be rectified and the test must be repeated as well as any others that may have been influenced by the fault.

612.1 General

... If any test indicates a failure to comply, that test and any preceding test, the results of which may have been influenced by the fault indicated, shall be repeated after the fault has been rectified ...

45 Which of the following is not a requirement for periodic inspection and testing?

○ a   Ensuring the safety of persons and livestock from shock and burns.
○ b   Ensuring the property is protected against damage from fire and heat due to an installation fault.
○ c   Confirmation that the installation is not damaged as to impair safety.
◉ d   Confirmation that the supply to the installation meets all the requirements of the Electricity Safety, Quality and Continuity Regulations.

Answer d

Options a, b and c are listed in Regulation 621.2. Option d is not a requirement. It is the responsibility of the Distribution Network Operator (DNO) to confirm full compliance of the supply.

621.2 Periodic inspection comprising a detailed examination of the installation shall be carried out without dismantling, or with partial dismantling as required, supplemented by appropriate tests from Chapter 61 to show that the requirements for disconnection times, as set out in Chapter 41 for protective devices, are complied with, to provide for:

(i)   **safety of persons and livestock against the effects of electric shock and burns**

(ii)   **protection against damage to property by fire and heat arising from an installation defect**

Notes

(iii) confirmation that the installation is not damaged or deteriorated so as to impair safety

(iv) the identification of installation defects and departures from the requirements of these Regulations that may give rise to danger.

NOTE: a generic list of examples of items requiring inspection is given in Appendix 6.

Section 7

46 For an electric floor heating system in a bathroom, a fine mesh metallic grid is <u>not</u> required to be connected to the protective conductor of the supply circuit if the protective measure used is

- ◉ a   SELV
- ○ b   electrical separation
- ○ c   a non-conducting location
- ○ d   an earth-free equipotential zone.

Answer a

Option b is not permitted in these circumstances (see Regulation 701.753). Options c and d are only for use where the installation is controlled or under the supervision of skilled or instructed persons, and so would not be permitted in a bathroom (see Regulations 701.410.3.6, 418.1 and 418.2).

701.753 Electric floor heating systems

For electric floor heating systems, only heating cables according to relevant product standards or thin sheet flexible heating elements according to the relevant equipment standard shall be erected provided that they have either a metal sheath or a metal enclosure or a fine mesh metallic grid. The fine mesh metallic grid, metal sheath or metal enclosure shall be connected to the protective conductor of the supply circuit. **Compliance with the latter requirement is not required if the protective measure SELV is provided for the floor heating system.**

For electric floor heating systems the protective measure "protection by electrical separation" is not permitted.

47 Access to the space under a bath is gained by using a screwdriver to remove the bath panel retaining clips. This space is classified as

- ○ a   zone 0
- ○ b   zone 1
- ◉ c   outside the zones
- ○ d   zone 2.

Answer c

A screwdriver (tool) is required for the panel removal, making the space outside the zones. See Regulation 701.32.3.

701.32.3 Description of zone 1

Zone 1 is limited by:

(i)   the finished floor level and the horizontal plane corresponding to the highest fixed shower head or water outlet or the horizontal plane lying 2.25 m above the finished floor level, whichever is higher

(ii)   the vertical surface:

    a)   circumscribing the bath tub or shower basin (see Figure 701.1)

    b)   at a distance of 1.20 m from the centre point of the fixed water outlet on the wall or ceiling for showers without a basin (see Figure 701.2 (e) and (f)).

Zone 1 does not include zone 0.

The space under the bath tub or shower basin is considered to be zone 1. **However, if the space under the bath tub or shower basin is only accessible with a tool, it is considered to be outside the zones.**

48 For SELV and PELV circuits in a bathroom, basic protection for equipment in all zones shall be provided by

- ○ a   obstacles
- ◉ b   barriers or enclosures
- ○ c   non-conducting location
- ○ d   placing out of reach.

Answer b

See Regulation 701.414.4.5, item (ii), which shows that option b is correct. Options a, c and d are not permitted in a bathroom (see Regulations 701.410.3.5 and 701.410.3.6).

701.414.4.5 Requirements for SELV and PELV circuits

Where SELV or PELV is used, whatever the nominal voltage, basic protection for equipment in zones 0, 1 and 2 shall be provided by:
(i)   basic insulation complying with 416.1, or
(ii)   barriers or enclosures complying with 416.2.

49 Electrical equipment in zones 0 and 1 of a fountain shall have mechanical protection to a minimum classification of

- ○ a   AG0
- ○ b   AG1
- ◉ c   AG2
- ○ d   AG3.

Answer c

In Regulation 702.55.3 the impact severity is given as medium, ie AG2. Option a does not exist and options b and d are too low and too high respectively. See Appendix 5 for definitions of AG1–3.

702.55.3 Electrical equipment of fountains

Electrical equipment in zones 0 or 1 shall be provided with mechanical protection to medium severity (AG2), eg by use of mesh glass or by grids which can only be removed by the use of a tool.

A luminaire installed in zones 0 or 1 shall be fixed and shall comply with BS EN 60598-2-18.

An electric pump shall comply with the requirements of BS EN 60335-2-41.

50 Which of the following would <u>not</u> be a requirement for an Assembly for Construction Sites (ACS), supplying mobile current-using equipment on a construction site?

- ◉ a   Main equipotential bonding.
- ○ b   Overcurrent protective devices.
- ○ c   Devices for fault protection.
- ○ d   Socket outlets.

Answer a

Options b, c and d are all requirements of Regulation 704.537.2.2. Option a is a process, not a device or accessory.

704.537.2.2 Devices for isolation

Each Assembly for Construction Sites (ACS) shall incorporate suitable devices for the switching and isolation of the incoming supply.

A device for isolating the incoming supply shall be suitable for securing in the off position (see Regulation 537.2.1.5), for example, by padlock or location of the device inside a lockable enclosure.

Current-using equipment shall by supplied by ACSs, each ACS comprising:

(i)   overcurrent protective devices, and

(ii)   devices affording fault protection, and

(iii) socket-outlets, if required.

Safety and standby supplies shall be connected by means of devices arranged to prevent interconnection of the different supplies.

---

51 **Which one of the following is <u>not</u> included in the scope of Section 705?**

○ a   Chicken-houses.

○ b   Stables.

○ c   Piggeries.

◉ d   Staff residences.

Answer d

Section 705 covers 'Agricultural and Horticultural Premises' and the scope is given in the first regulation. Staff premises are excluded, where the standard requirements apply.

705.1 Scope

The particular requirements of this section apply to fixed electrical installations indoors and outdoors in agricultural and horticultural premises. Some of the requirements are also applicable to other locations that are in common buildings belonging to the agricultural and horticultural premises. Where the special requirements also apply to residences and other locations in such common buildings this is stated in the text of the relevant regulations.

Rooms, locations and areas for household applications and similar are not covered by this section.

Notes

52 In conducting locations with restricted movement, supplies to handlamps shall be provided by

- ⊙ a   SELV
- ○ b   reduced low voltage supplies
- ○ c   RCD or RCBO
- ○ d   PELV.

Answer a

The requirements for handlamps in restrictive conductive locations are given in Regulation 706.410.3.10 , item (ii).

Reduced low voltage and RCDs (options b and c) are not allowed. PELV (option d) is only permitted for the supply to fixed equipment.

53 The minimum height of overhead conductors on caravan sites, when installed in vehicle movement areas, is

- ○ a   3.0 m
- ○ b   3.5 m
- ○ c   5.8 m
- ⊙ d   6.0 m

Answer d

The requirements for electrical installations in caravan parks are given in Section 708. Where overhead conductors are installed they must be at least 6 m high in vehicle movement areas; see Regulation 708.521.1.2.

708.521.1.2 Overhead distribution circuits
All overhead conductors shall be insulated.

Poles and other supports for overhead wiring shall be located or protected so that they are unlikely to be damaged by any foreseeable vehicle movement.

Overhead conductors shall be at a height above ground of not less than 6 m in all areas subject to vehicle movement and 3.5 m in all other areas.

54 The particular requirements for mobile or transportable units apply to

- ○ a   generating sets
- ◉ b   mobile workshops
- ○ c   mobile machinery to BS EN 60204-1
- ○ d   traction equipment of electric vehicles.

Answer b
See Regulation 717.1, which shows that options a, c and d are not applicable.

717.1 Scope
The particular requirements of this section apply to a.c. and d.c. installations for mobile or transportable units.

For the purposes of this section, the term "unit" is intended to mean a vehicle and/or mobile or transportable structure in which all or part of an electrical installation is contained, which is provided with a temporary supply by means of, for example, a plug and socket-outlet.

Units are either:
(i)   of the mobile type, eg vehicles (self-propelled or towed), or
(ii)   of the transportable type, eg containers or cabins.
Examples of the units include technical and facilities vehicles for the entertainment industry, medical or health screening services, welfare units, promotion and demonstration, fire fighting, **workshops**, offices, transportable catering units etc.

The requirements are not applicable to:
(iii)   generating sets
(iv)   marinas and pleasure craft
(v)   mobile machinery in accordance with BS EN 60204-1
(vi)   caravans to Section 721
(vii) traction equipment of electric vehicles
(viii)   electrical equipment required by a vehicle to allow it to be driven safely or used on the highway.
Additional requirements shall be applied where necessary for units including showers, or for medical locations, etc.

55 Electrical equipment in a circus installation must have a degree of protection of at least

○ a   IP33
○ b   IP4X
◉ c   IP44
○ d   IPX8.

Answer c

The requirements for circuses are given in Section 740. Regulation 740.512.2 indicates that electrical equipment in a circus must have a degree of protection of at least IP44. This is similar to that for agricultural locations where there is also livestock.

740.512.2 External influences

Electrical equipment shall have a degree of protection of at least IP44.

Section 8

56 The ambient air temperature rating factor for 90 °C thermosetting cables operating in an ambient air temperature of 60 °C is

○ a   0.50
○ b   0.56
○ c   0.63
◉ d   0.71.

Answer d

Correction factors for ambient air temperature are given in Table 4B1 of Appendix 4.

Table 4B1 Rating factors (Ca) for ambient air temperatures other than 30 °C

Notes

Ambient temperature °C	Insulation				
	60 °C thermosetting	70 °C thermoplastic	**90 °C thermosetting**	Mineral	
				Thermoplastic covered or bare and exposed to touch 70 °C	Bare and not exposed to touch 105 °C
25	1.04	1.03	1.02	1.07	1.04
30	1.00	1.00	1.00	1.00	1.00
35	0.91	0.94	0.96	0.93	0.96
40	0.82	0.87	0.91	0.85	0.92
45	0.71	0.79	0.87	0.78	0.88
50	0.58	0.71	0.82	0.67	0.84
55	0.41	0.61	0.76	0.57	0.80
60		0.50	**0.71**	0.45	0.75

57 A BS 3036 100 A fuse, when carrying 1800 A, has a disconnection time of

From the *IET Wiring Regulations*, extract from Table 4B1, page 325

- a   0.2 second
- b   0.4 second
- c   4 seconds
- d   5 seconds.

Answer a
Overcurrent device characteristics are given in Appendix 3. Figures 3A2 (a) and (b) give data for BS 3036 fuses (see Figure 3A2 (b), on the following page). When carrying 1800 A, a 100 A fuse will have a disconnection time of 0.2 seconds.

Notes

Fig 3A2(b) – Semi-enclosed fuses to BS 3036

Fuse rating	Maximum operating time/current characteristics for semi-enclosed fuses to BS 3036				
	Current for time				
	0.1 sec	0.2 sec	0.4 sec	1 sec	5 secs
20 A	260 A	180 A	130 A	92 A	60 A
45 A	900 A	580 A	390 A	250 A	145 A
100 A	2800 A	1800 A	1200 A	770 A	430 A

From the *IET Wiring Regulations*,
Figure 3A2(b), page 298

58 A 2.5 mm² thermoplastic insulated and sheathed flat cable with protective conductor is laid in a ceiling beneath thermal insulation 80 mm thick in contact with the ceiling board, as shown in the figure below. What is its installed rating?

- ○ a   17 A.
- ○ b   20 A.
- ◉ c   21 A.
- ○ d   27 A.

Answer c

Refer to Table 4A2 of Appendix 4. Reference method 100 is appropriate in this instance.

Refer then to Table 4D5, which gives the cable rating for reference method 100 and 2.5 mm² cable as 21A.

59 The maximum trip time for a general purpose, non-delay 100 mA RCD to BS EN 61008-1 when subjected to a current of 5 times its rating is

- ◉ a   40 ms
- ○ b   150 ms
- ○ c   200 ms
- ○ d   300 ms.

Answer a

Table 3A in Appendix 3 gives time/current performances for various types and ratings of RCD and shows that all general purpose RCDs should trip within 40 ms at 5 times their rating.

Notes

Table 3A – Time/current performance criteria for RCDs to BS EN 61008-1 and BS EN 61009-1

RCD type	Rated residual operating current $I_{\Delta n}$ mA	Residual current mA	Trip time ms	Residual current mA	Trip time ms	Residual current mA	Trip time ms
General non-delay	10	10	300 max.	20	150 max.	50	**40 max.**
	30	30		60		150	
	100	100		200		**500**	
	300	300		600		1500	
	500	500		1000		2500	
Delay 'S'	100	100	130 min.	200	60 min.	500	40 min.
	300	300	500 max.	600	200 max.	1500	150 max.
	500	500		1000		2500	

60 The installation reference method for a cable installed under a ceiling is

○ a   Method A
◉ b   Method B
○ c   Method C
○ d   Method D.

Answer b

Option c would seem to be the logical choice, but Section 7.2 in Appendix 4 gives the reason that option b is the correct answer.

7.2 Other Methods

Cable under a ceiling: This installation may appear similar to Reference Method C but because of the reduction of natural air convection, Reference Method B is to be used for the current rating.

Answer key

Sample test 2

Question	Answer	Question	Answer
1	a	31	c
2	c	32	a
3	b	33	a
4	b	34	a
5	d	35	a
6	c	36	d
7	b	37	c
8	a	38	b
9	b	39	b
10	d	40	c
11	c	41	d
12	b	42	a
13	d	43	b
14	a	44	c
15	a	45	d
16	b	46	a
17	b	47	c
18	b	48	b
19	b	49	c
20	a	50	a
21	d	51	d
22	d	52	a
23	b	53	d
24	b	54	b
25	b	55	c
26	d	56	d
27	d	57	a
28	c	58	c
29	b	59	a
30	d	60	b

Notes

Notes

More information

More information

Further reading

Required reading
BS 7671:2008(2011) Requirements for Electrical Installations (IET Wiring Regulations, Seventeenth Edition, incorporating Amendment No 1), published by The IET, London, 2011

Additional reading
On-Site Guide (BS 7671:2008 Wiring Regulations, incorporating Amendment No 1: 2011), published by The IET, London, 2011

The Electrician's Guide to Good Electrical Practice, published by Amicus, 2008

Electrician's Guide to the Building Regulations, published by The IET, London, 2008

IET Guidance Notes, a series of guidance notes, each of which enlarges upon and amplifies the particular requirements of a part of the Wiring Regulations, published by The IET, London:
– Guidance Note 1: *Selection and Erection of Equipment*
– Guidance Note 2: *Isolation and Switching*
– Guidance Note 3: *Inspection and Testing*
– Guidance Note 4: *Protection Against Fire*
– Guidance Note 5: *Protection Against Electric Shock*
– Guidance Note 6: *Protection Against Overcurrent*
– Guidance Note 7: *Special Locations*
– Guidance Note 8: *Earthing and Bonding*

Brian Scaddan, *Electrical Installation Work*, published by Newnes, 2011

John Whitfield, *Electrical Craft Principles*, published by The IET, London, 2009

Online resources

City & Guilds www.cityandguilds.com
The City & Guilds website can give you more information about studying for further professional and vocational qualifications to advance your personal or career development, as well as locations of centres that provide the courses.

Institution of Engineering and Technology (IET) www.theiet.org
The Institution of Engineering and Technology was formed by the amalgamation of the Institution of Electrical Engineers (IEE) and the Institution of Incorporated Engineers (IIE). It is the largest professional engineering society in Europe and the second largest of its type in the world. The Institution produces the *IET Wiring Regulations* and a range of supporting material and courses.

SmartScreen www.smartscreen.co.uk
City & Guilds' dedicated online support portal SmartScreen provides learner and tutor support for over 100 City & Guilds qualifications. It helps engage learners in the excitement of learning and enables tutors to free up more time to do what they love the most – teach!

British Standards Institution www.bsi-global.com

Gas Safe Register www.gassaferegister.co.uk

ELECSA Ltd www.elecsa.org.uk

Electrical Contractors' Association (ECA) www.eca.co.uk

Joint Industry Board for the Electrical Contracting Industry (JIB)
www.jib.org.uk

NAPIT Certification Services Ltd www.napit.org.uk

National Inspection Council for Electrical Installation Contracting (NICEIC) www.niceic.com

Oil Firing Technical Association for the Petroleum Industry (OFTEC)
www.oftec.co.uk

Further courses

City & Guilds Level 3 Award in the Initial Verification and Certification of Electrical Installations (2394)

This qualification helps candidates to develop the knowledge and practical skills needed to professionally design and install or inspect and test electrical installations. The Award is ideal for qualified electricians wishing to update their knowledge of electrical installation techniques and legislation.

City & Guilds Level 3 Award in the Periodic Inspection, Testing and Certification of Electrical Installations (2395)

This Award will enable candidates to update their knowledge and skills in electrical installation and will boost their confidence and ability in carrying out inspections and testing. This Award has been developed for practising electricians who have not carried out inspection and testing since qualifying or who require some update of training before progressing to other City & Guilds qualifications.

City & Guilds Level 4 Award in the Design and Verification of Electrical Installations (2396)

This qualification helps to develop the essential, up-to-date knowledge and techniques needed to professionally design and install or inspect and test electrical systems. This Award is ideal for qualified electricians wishing to expand or update their professional knowledge and skills.

City & Guilds Level 3 Certificate in the Code of Practice for In-service Inspection and Testing of Electrical Equipment (2377)

This course, commonly known as PAT/Portable Appliance Testing, is for staff undertaking and recording inspection and testing of electrical equipment. The course includes a practical exercise. Topics covered: equipment construction, inspection and recording, combined inspection and testing, and equipment.

City & Guilds Level 2 Certificate in Fundamental Inspection, Testing and Initial Verification (2392-10)

This qualification was developed to meet industry needs and to provide candidates with an introduction to the initial verification of electrical installations. It is aimed at practising electricians who have not carried out inspection and testing since qualifying, those who require update training and those with limited experience of inspection and testing. Together with suitable on-site experience, it would also prepare candidates to go on to the Level 3 Certificate in Inspection, Testing and Certification of Electrical Installations (2392-20).

City & Guilds Building Regulations for Electrical Safety

This new suite of qualifications is for Competent Persons in Domestic Electrical Installations (Part P of the Building Regulations). The qualifications consist of components for specialised domestic building regulations and domestic wiring regulations routes as well as a component for Qualified Supervisors.

JIB Electrotechnical Certification Scheme (ECS) Health and Safety Assessment

This Health and Safety Assessment is a requirement for electricians wishing to work on larger construction projects and sites in the UK and the exam is online. It is now a mandatory requirement for holding an ECS card, and is a requirement for all members of the ECS. Please refer to www.jib.org.uk/ecs2.htm for details.

Notes